Janaína Accordi Junkes
Ana Maria Segadães
Dachamir Hotza

Formulating ceramics from mineral waste

Janaína Accordi Junkes
Ana Maria Segadães
Dachamir Hotza

Formulating ceramics from mineral waste

Using a phase diagram as a guide

ScienciaScripts

Imprint

Cover image: www.ingimage.com

This book is a translation from the original published under ISBN 978-3-330-76876-5.

Publisher:
Sciencia Scripts
is a trademark of
Dodo Books Indian Ocean Ltd. and OmniScriptum S.R.L publishing group

120 High Road, East Finchley, London, N2 9ED, United Kingdom
Str. Armeneasca 28/1, office 1, Chisinau MD-2012, Republic of Moldova, Europe
Managing Directors: Ieva Konstantinova, Victoria Ursu
info@omniscriptum.com

Printed at: see last page
ISBN: 978-620-8-60047-1

SUMMARY

PRESENTATION

In recent years, scientific research has shown that environmental preservation issues have become very important and that one of the major challenges to be met is the recycling of materials discarded by different productive sectors. Due to the damage caused to the environment by technological development through the disposal of waste, this study sought to evaluate the possibility of using only industrial waste as alternative raw materials in the manufacture of ceramic tiles. Different industrial wastes classified as non-hazardous were selected: sludge from the gneiss crushing process, sludge from the varvite cutting and polishing process, sludge from the drinking water clarification/purification process and a clay also classified as waste. All the residues, as they were generated, were dried and disaggregated in bubble mills, and characterized by X-ray fluorescence, X-ray diffractometry, differential and gravimetric thermal analysis, optical dilatometry and particle size distribution. The applicability of these residues in the manufacture of rustic ceramics for coating was guided by the phase diagram of the SiO_2-K_2O-$Al_{(2)}O_3$ system and four formulations were established. For the initial tests, these formulations were homogenized and pressed into tablets and sintered at 900, 950, 1000, 1050, 1100 and 1150°C. The plasticity of the formulations was assessed using the Casagrande method, and the extrusion technique was used as the forming process. The extruded samples were sintered at 1100 and 1150°C with 40 min of setting and characterized by X-ray diffractometry, differential and gravimetric thermal analysis, optical dilatometry, linear shrinkage, water absorption and mechanical flexural strength. The crystalline phases identified were associated with the firing conditions (temperature, time, atmosphere), as well as the intrinsic characteristics of the raw materials, such as chemical composition, particle size distribution and homogeneity. The residues proved to be a good source of alternative raw materials and their corresponding formulations proved feasible for the manufacture of ceramic tiles.

1 INTRODUCTION

One of the most important problems that has accompanied humanity in recent decades is environmental degradation. The exploitation of natural resources and technological development are events which, if left unchecked, can cause ecological imbalances with unpredictable consequences. In this context, the recycling of waste is of major importance.

Recycling, understood here as the reuse of waste from a process, is more widely used in countries where raw materials have strategic aspects and solid waste disposal techniques have considerable economic costs. Ecological awareness is beginning to be awakened and, in recent years, stricter environmental control laws have been implemented, in general, and licensing of waste-generating industrial activities, in particular.

Environmental legislation determines that the generator is responsible for the waste, from its generation to its final destination, imposing administrative, civil and criminal liability for damage caused to humans and the environment as a result of non-environmentally appropriate management of solid industrial waste.

Among the advantages of recycling waste, regardless of its type, compared to using "virgin" natural resources, are a reduction in the volume of raw materials extracted, a reduction in energy consumption, lower emissions of pollutants and an improvement in the health and safety of the population. Preserving natural resources, extending their useful life and reducing environmental destruction is the most visible advantage of recycling.

The accelerated process of industrialization observed in some countries, combined with demographic expansion, has led to a considerable increase in the production of solid waste, particularly industrial waste. The inadequate treatment of solid industrial waste contributes significantly to the worsening of environmental problems.

All over the world, research into the recycling of industrial waste is being intensified. Recycling is seen by private enterprise in North America and Europe as a highly profitable market. Many companies invest in research and technology, which increases the quality of the recycled product and makes the production system more efficient.

In recent years, the reuse of various solid mineral residues has been growing. In addition to mineral extraction residues, residues from coal processing, the textile industry and the ceramics industry itself have been tested as raw materials for the manufacture of ceramic, vitreous and glass-ceramic products.

The ceramics industry began to recycle or reuse materials many years ago, starting with the use of breakages from the process, known as chamotte, which are ground up and added to new masses, thus reducing the amount of loss in the process and the costs of disposal. However, due to the high inertness of ceramics, this sector has become a target for recycling materials from other industries.

Over the last 30 or 40 years, awareness has been growing, particularly in highly industrialized countries, of the potentially devastating effects on the environment of hazardous and toxic waste, the production of which is increasing. The vitrification process, initially proposed for higher radioactive levels in waste management, has proven to be one of the most valid technologies for inertizing and reducing the volume of different categories of waste. The glass produced can in fact be subsequently disposed of in landfills without any danger to the environment [Colombo et al., 2003].

Due to the heterogeneity of traditional ceramic products, they allow a reasonable amount of waste to be incorporated without damaging the properties of the end products, thus reducing the amount of waste. Because it is a large-scale production, the ceramics industry consumes a significant amount of waste, even if it absorbs small quantities.

The ceramics industry encompasses a very wide range of compounds, so a multitude of products can be made using different methods. Therefore, this industry presents favorable conditions for the implementation of waste recovery systems.

Making products from waste is an advantage that puts the manufacturer in a highly competitive position on the market, due to the economic issue involved and the opportunity to market this principle, especially with regard to the ecological aspect. It is in this context that the use of the phase diagram becomes a useful tool for guiding ceramic production and also helping to make good choices about composition and processing parameters.

An understanding of phase equilibrium in ceramic systems can be very useful in the use and development of refractory materials, glasses and other products made at high temperatures. In the processing and manufacture of ceramic products, the reactions that take place are more clearly understood if the relationship between the phases under equilibrium conditions is known. The physical and chemical properties of ceramic products are related to the number, composition and distribution of the phases present. Temperature, pressure and concentration are the main variables that determine the types and quantities of phases present under equilibrium conditions [Bergeron and Risbud, 1984].

1.1. GENERAL OBJECTIVE

The aim of this work is to formulate a new material, produced exclusively from waste, for use as a rustic ceramic coating, using the phase diagram to guide its production.

1.2. SPECIFIC OBJECTIVES

In order to achieve the general objective, considering the technological innovation involved in the processing of ceramic tiles using only mineral residues, it is necessary to highlight some items that underpin and guide the various stages of the production process. To this end, the following specific objectives have been defined:

- Check which mineral residues are available in the state of Santa Catarina;
- To evaluate the potential use of these residues as alternative raw materials for the ceramics industry;
- Characterize the alternative raw materials selected;
- Identify the phase diagram(s) and their relevant regions for the formulation of ceramic products taking into account the selected residues;

- Determine, using the phase diagram, the percentage of residue to be used in the formulations;
- Selecting the best process parameters with the help of the phase diagram;
- Test the chosen parameters; and
- Evaluate the results obtained through tests and analysis.

1.3. ORIGINAL CONTRIBUTION TO THE BOOK

With regard to the original contributions of the work carried out to the field of ceramic materials, the following can be mentioned:

- To present the possibility of reusing mineral waste produced in the state of Santa Catarina by using it to produce structural ceramic materials, with the aim of adding value to the waste and contributing to sustainable development;
- Use a ternary phase diagram to help choose compositions and heat treatment and to predict the tendency for the reaction to be completed in ceramic materials developed only from Santa Catarina mineral waste.

1.4. BOOK STRUCTURE

This book is divided into five chapters organized as follows:

Chapter 1 presents a brief contextualization of the topic, the objective and a description of the organization of the book.

Chapter 2 deals with aspects related to the recycling of industrial waste, as well as the ceramic technology involved in this study, covering definitions and classifications of each, as well as information on the process of formation of each waste selected.

Chapter 3 describes the methodology used in this study, covering all the processing stages, including selection, processing and characterization of the raw materials, definition of the formulations and obtaining the samples.

Chapter 4 refers to the results and discussions. This chapter presents the characterization of the waste, as well as the results of the tests applied to the established formulations.

Finally, Chapter 5 presents the conclusions of all the work that has been done. This chapter also suggests some topics for further study that could complement this work.

2. THEORETICAL BASIS

2.1. solid waste and its recovery

Everything that surrounds us - such as houses, cars, bridges, planes - will one day be waste. To this total must be added all the waste from the process of extracting raw materials and producing goods. Thus, in any society, the amount of waste generated exceeds the amount of goods consumed.

Economic and demographic growth, with the adoption of ever higher and healthier standards of living, demands continued industrial activity. Such activity brings with it two environmental problems: firstly, the consumption of non-renewable resources, making them increasingly scarce and eventually leading to their extinction; and secondly, an increase in the amount of waste, which is increasingly difficult to dispose of. In this way, industry faces two new concerns: natural raw materials are becoming more and more expensive, and waste is becoming a burden with ever higher disposal costs. In general, all types of industries have begun to look for alternative, less expensive solutions to natural raw materials, and have also optimized their processes to make their products as much as possible, with the simultaneous advantage of producing less waste [Segadaes, 2006].

The concern of society and industry with environmental issues has awakened in companies the idea that pollution is synonymous with waste and that waste is not garbage, thus changing the inadequate concept that waste is useless or disposable material.

Concern about the environment in Brazil triggered the formulation of a new law (No. 12.305, of August 2, 2010) establishing the national solid waste policy. The government's new guidelines establish actions by which manufacturers, importers, distributors and traders of batteries, pesticides, tires, electrical and electronic products and their components, among others, are obliged to structure and implement reverse logistics systems, by returning the products after use by the consumer [Legislation 2010].

This law (article 3, section XVI) defines solid waste as discarded material or goods resulting from human activities in society, the final destination of which is solid or semi-solid [Legislaçao 2010].

The term waste is normally used to refer to what is left over from a production process; when there is no application for it, it is called a reject. However, when waste can be incorporated as a raw material or energy source in the production of new materials, it is classified as a by-product.

Traditionally, waste is disposed of in specific industrial landfills, which can lead to other problems such as the rapid depletion of landfill capacity, the existence of leachates that are difficult to treat, or gaseous emissions. According to Rocha and John [2003], the costs of disposing of waste in landfills include packaging, treatment, transportation and environmental licensing. Therefore, from an ecological and also economic point of view, recycling is a way of avoiding the inconvenience that disposing of or storing waste causes the community and the companies that generate it. In addition to the direct costs, there are indirect costs, such as the erosion of the company's image due to its inefficient environmental management, which can lead to confrontations with social organizations and the loss of consumers. These factors can determine interest in recycling technology.

Recycling waste is nothing new, particularly in industries that use large amounts of energy, such as steel and iron, aluminum, glass and paper. When it comes to consuming or reusing other industrial waste, the ceramics industry, particularly the sector dedicated to the manufacture of building materials, is at the top of the list. Natural raw materials used in the manufacture of clay-based ceramic products vary widely in composition and the resulting products are very heterogeneous [Segadaes, 2006]. For this reason, such products can tolerate greater fluctuation in their composition and changes in raw materials, and the ceramics industry is able to incorporate a variety of residues [Raupp-Pereira et al., 2004]. Even if this process is carried out in small quantities, the high production rates will translate into a significant consumption of waste. Furthermore, given the reasonably high firing temperature generally used (around 1000°C), the effective incorporation of materials within a ceramic matrix is achieved, which is particularly attractive when it comes to the inertization of toxic waste [Segadaes et al., 2005].

Inertization is the process used to slow down the release of hazardous substances into the environment. The technique consists of incorporating the waste into a matrix, which in some cases may require pre-treatment of the waste in order to adapt the physical or chemical characteristics of the waste to the technical needs of the process. However, over time, some release may occur and instead of complete inertization, which is considered a definitive retention, it is more correct to assume that the substances are immobilized and controlling the rate of release is the key to controlling the problems [Conner, 1992].

The inertization or stabilization process according to LaGrega et al. [1994] can involve the following chemical and physical mechanisms: (i) macro-encapsulation, (ii) micro-encapsulation, (iii) adsorption, (iv) absorption, (v) precipitation. The first two mechanisms are the most common for incorporation into a clay ceramic matrix. The simple formation of a physical barrier which reduces leachability is a typical macro-encapsulation process.

For waste to be used in industry, the quantity of waste produced must be significant, as the low quantity of waste generated limits the commercial recycling options that require industrial processing, which is always sensitive to scale [Rocha and John, 2003].

According to Pinto [2004], there are important basic requirements for the reuse of waste in the ceramics industry:

- Selective and temporary storage, in order to guarantee the obtaining of batches of significant size and free of other types of waste,
- Mixing of different batches, with the aim of ensuring good homogeneity and
- Pre-treatment, where necessary, usually involving simple operations such as grinding, drying, de-fermentation and calcination.

The competitive advantage of the new product is very important, as it is a decisive factor in its success on the market. Competitive advantage can be understood as that which reduces the relative cost of the product to a product of the same performance or which results in a higher performance product. To attract the interest of the waste generator from a strictly financial point of view, recycling needs to reduce waste costs, including costs arising from the need to change the treatment of the waste in order to make it suitable for recycling. Overall, the investment made in recycling must offer an attractive rate of return [Rocha and John, 2003].

Recycling industrial waste as a raw material in the development of products in the ceramics sector:

- Identification and quantification of available waste;
- Physico-chemical and microstructure characterization of the waste, including its environmental risk;
- Search for possible applications in the ceramics sector, considering the characteristics of the waste;
- Analysis of costs associated with waste;
- Development of different applications, including their production process;
- Performance analysis in relation to different user needs for each specific application;
- Analysis of the environmental impact of the new product, in an approach that must necessarily involve an assessment of the risks to the health of workers and users.

The use of industrial waste in ceramic products or other types of materials can be an interesting solution, both from an environmental and economic point of view, since the use of waste means a reduction in the use of natural resources, no landfill and lower transportation costs.

Although it is possible and a priority to reduce the amount of waste during production and even post-consumption, it will always be generated and sustainable development requires a reduction in the consumption of non-renewable natural raw materials. Therefore, closing the production cycle, generating new products by recycling waste, is an irreplaceable alternative. Therefore, the development of environmentally efficient and safe waste recycling technologies that result in products with adequate technical performance and that are economically competitive in different markets is an important technical challenge [Rocha and John, 2003].

Solid waste can be generated as a by-product of industrial processes or as sludge from effluent treatment plants. It is important to know which operations generate waste, as well as the factors that affect its generation in manufacturing processes [Teixeira, 2002a].

2.1.1. Ceramic Materials Produced from Waste

Ceramics can be defined as inorganic, non-metallic materials that are typically produced using clays and other natural minerals and are obtained after heat treatment at high temperatures [ABCERAM, 2010].

The performance of ceramic products is linked to their chemical composition, which determines their properties. In the development and production of ceramics, strict control of materials and process operations is also required to minimize defects. The manufacture of ceramic materials is a complex interaction between raw materials, technological processes, people and financial investment [Reed, 1995].

Modern construction techniques, growing competitiveness in international markets and technological renewal demand products that meet the fundamental requirements of high quality and low cost. Knowledge of the raw materials and their effect on the processing stages and properties of the final product are necessary to meet these requirements.

In order to study ceramic processes, it is necessary to be familiar with the types of raw materials available. Argillominerals provide plasticity when mixed with water; feldspars or feldspathoids act as non-plasticizing agents in forming and form a liquid phase in sintering; silica (quartz) resists fusion and has been the backbone of traditional ceramics. Other mineral silicates are used in ceramic coatings, such as cordierite-based products, which are resistant to thermal shock, and steatite electrical porcelains [Reed, 1995].

It is often overlooked that some waste materials are similar in composition to the raw materials currently used, and often contain materials that are not only compatible, but also bring some benefit to the manufacture of ceramics [Santos, 1997]. Within this scenario, elevating industrial waste to the category of alternative raw materials becomes interesting, both technically and economically, for a wide range of applications, which could mean the much sought-after low-cost alternative raw material [Segadaes et al., 2005].

Several research projects have been dedicated to the development of new technologies aimed at reducing or eliminating the negative environmental impact caused by the industrial process, basically through the use of waste materials as alternative raw materials for the ceramics industry. Examples include the thermal and/or chemical extraction of high-purity silica from rice husks or rice husk ash, which can be used as a raw material in ceramics [Della et al., 2001, Della et al., 2002, Della et al., 2003, Della et al., 2005 and Junkes et al., 2006]. Alternatively, heavy ash from municipal solid waste incineration processes can be used in ceramic products such as porcelain stoneware, and to obtain frits with improved water absorption and stain resistance characteristics, without significantly altering flexural strength [Andreola, et al., 2001, Barbieri, et al., 2002 and Appendino, 2004].

Waste from different groups of materials, when combined, can also produce quality ceramic products. An example of such a combination is the development of a ceramic pigment using rice husks and steel mill scale as alternative raw materials. To obtain the red ceramic pigment, the rice husk is chemically and thermally treated to obtain amorphous silica and the steel mill scale is thermally treated to obtain hematite with a reddish hue. The results obtained from this research confirm the possibility of valorizing and recycling rice husks, which are rich in silica, as a matrix and steel mill scale to obtain hematite, thus obtaining an encapsulated ceramic pigment [Della et al.,

2007].

The use of residues from the processing of ornamental stones such as marble and granite has already been explored by adding percentages of residue, which can reach up to 50%, to a clay. These residues can replace traditional fluxes, with the advantage of controlling the plasticity and shrinkage of the ceramic body without having a negative effect on the properties of the products, while also allowing sintering at lower temperatures than conventional ones, resulting in energy savings [Acchar et al., 2005, Acchar et al., 2006a, Acchar et al., 2006b, Segadaes et al., 2005].

Replacing natural raw materials with waste can be an opportunity to improve the current difficulties of waste management. Even if this is done in small quantities, high production rates can translate into significant consumption of waste and, for industries willing to use them, the latter can constitute cheap raw materials from renewable sources [Raupp-Pereira et al., 2007].

2.1.2. Valorization of Mineral Waste from Santa Catarina

2.1.2.1. Gneiss

One of the residues available for recovery in the state of Santa Catarina is gneiss mud. Teixeira et al. [2000] mention that "gneiss" refers to rocks made up predominantly of feldspars and quartz, with at least 20% feldspar in volume. Gneisses originating from granitic rocks are called orthogneisses; when they originate from sedimentary rocks they are called paragnaises. Gneisses can also originate from the progressive metamorphism of mica schists, when they are rich in quartz, and can contain garnet, cordierite, kyanite or sillimanite, or arkose (feldspathic sandstones) or grauvacite (sandy sedimentary rocks rich in clay minerals) [Silveira, 2005].

According to Valente and Barbosa [2003], gneiss is a rock widely used as gravel in construction and paving, as well as for ornamental purposes. This type of rock is suitable for crushing due to aspects such as its strength, low porosity and regular fragmentation.

Crushed stone comprises rock fragments in the particle size range of 100 mm and 6 mm. These crushed stones are mainly used in concrete for construction and asphalt for road surfacing. Asphalt is made up of 92% gravel and sand. In terms of material quality, crystalline rocks with high mechanical strength, low porosity, favorable mineral composition and no alteration minerals are used, for example granites, gneisses, basalts, quartzites and crystalline limestones [DRM, 2009].

Gneiss waste has already been used in the ceramics industry with good results, as this rock has a predominant mineral composition of quartz, feldspars and mica. In the stages prior to firing, these minerals act as inert agents, reducing the amount of water needed to shape the pieces and facilitating drying. During firing, quartz acts as an inert and may also be dissolved in the liquid phase formed and, depending on its particle size and temperature, react with other constituents. Feldspars and mica favor the formation of the liquid phase, helping to reduce the porosity of the pieces [Vieira et al., 2006a]. The fineness and low cost of the residue make it technically and economically viable for use in ceramics.

In their study, D'Agostino and Soares [2003] showed the possibility of using the fines resulting from crushing granite-gneiss rock to prepare mortars. This material had a good particle size distribution, allowing for better filling of voids left by larger particles and making it possible to prepare mortars with smaller amounts of cement. The granite-gneiss rock fines were completely surrounded by the cement paste, thus providing a better bond between aggregate and binder and, consequently, greater mortar strength.

The residue from the crushing and polishing of gneissic rock is a material that can be used in the composition of electrical porcelain, due to its mineralogical composition, with the presence of feldspathic and quartz fluxes. Mixing this residue with clay allows the development of a microstructure after sintering similar to traditional porcelain compositions [Vieira et al., 2006b].

Experiments carried out with ceramic bodies containing up to 20% by mass of gneissic ornamental rock fines showed good technological properties in terms of linear shrinkage, water absorption and mechanical strength. Furthermore, the incorporation of waste does not alter the red ceramic production process [Moreira et al., 2005,

Moreira et al., 2008].

Vitrified sidewalks containing up to 47.5% by mass of gneiss rock fines were prepared and evaluated by Souza et al. [2010]. The residue used is a low-cost material, rich in potassium oxide (K_2O), which can replace conventional fluxes (sodium and potassium feldspars) in sidewalk formulations. The results showed that samples from the BIIb group were *upgraded* to the BIa group in terms of water absorption and mechanical strength according to the ISO 13006 standard, showing that recycling gneissic rock fines in vitrified sidewalks can be a technological solution with many economic and environmental benefits.

2.1.2.2. Varvito

Another rock that produces sufficient residue from its processing and is available in Santa Catarina for recovery is varvite. This rock of sedimentary origin is formed by the repeated succession of sheets or layers deposited over a period of years. Each varvite sheet or layer is actually a pair made up of a thicker (in the order of cm) lower section of sandstone or siltstone, which is lighter in color, followed by a thinner (in the order of mm) section of siltstone or claystone, which is darker in color (dark gray). The term varvito derives from the Swedish word *varve*, generally used to refer to seasonal sedimentary deposits, i.e. those controlled by the variations in the seasons [Souza, 2008].

Varvitic deposits represent bottom sediments in glacial lakes. Their structure is characterized by parallelism and homogeneity of the layers. They show perfect cyclicity characterized by lighter, coarser layers interspersed with thinner, darker layers with a high incidence of organic matter. The coarser varve represents the thawing seasons when the lake receives a lot of sediment. The finer, darker varve represents the cold season when no sediment reaches the lake and there is precipitation of clays and organic matter, resulting from the death of microorganisms [Popp, 1998].

Varvite sludge is generated by the use of water during cutting which, when mixed with the varvite powder (fines), forms a sludge. Another source of waste, also considered fines, is the polishing of cut pieces. Despite the large volume of material mined and removed from quarries, in some companies the process of extracting and processing the rock is still done in a rudimentary way, using trolleys with small diamond blades for cutting and guides to obtain the dimensions of the desired piece, such as floors, table tops, benches, etc. [Cunha, 2007].

All the waste from the mining and exploitation of varvito creates major technological and environmental problems, as it accumulates around the mine, creating huge mountains of waste. To avoid problems during the extraction of the rock, a large part of the waste is transported to locations close to the quarries, causing great damage to the region's fauna and flora. The estimated quantity of this waste ore is in the order of 50-60 thousand tons/month, in just one of the mines located in the town of Trombudo Central [Cunha, 2007].

Varvito waste has already been used to make ceramic pieces by bonding. The characterization of the rock powder shows that the quartz concentration phase is high and would be sufficient to produce ceramic pieces. Optimizing the properties of the pieces for different uses can be achieved by controlling the characteristics of the powder, the moulding process and the sintering conditions [Souza et al., 2004; Mansur et al., 2006].

Campos et al. [2004] used powder technology to manufacture floor tiles using varvito fines, studying the best compaction pressure and sintering temperature to achieve the best properties, as well as the maximum particle size of the fines, for good quality tiles. The results obtained by the authors proved that varvite fines are suitable for use as a raw material for ceramic tiles, *since* their properties are within the standards used to make conventional ceramics.

Silva et al. [2006] investigated the products resulting from the expansion process of varvite mining fines as possible aggregates in the manufacture of pozzolanic cement. The pozzolanic activity of the products obtained by expansion was confirmed by the results of the mechanical strength of the mortars. Varvite has no pozzolanic activity, but its chemical composition is compatible with use as a mineral additive to clinker, since the fines that have undergone expansion can partially replace Portland cement clinker as a pozzolanic additive.

Powders resulting from cutting and machining in varvite quarries have also been processed by pressing and sintering for the manufacture of ceramic products, such as flooring and tiles. An important point to take into account is the particle size of the raw materials, which is why the atomization process was used to homogenize the particle sizes and ensure adequate feeding of the pressing systems. The results obtained showed that the use of varvite fines made it possible to produce products with superior properties when compared to traditional ceramics. Furthermore, by using this fines as a raw material, it is possible to solve two problems: the elimination of large quantities of waste and the improvement of the mechanical performance of common roof tiles used in civil construction [Catarino et al., 2003].

2.1.2.3. Water Treatment Plant Sludge

A waste product widely used in research and available in Santa Catarina is sludge from drinking water filtration/clarification processes. The vast majority of drinking water that reaches a home ends up as sewage, which is reintroduced into rivers and lakes. Once contaminated, these water sources can contain microorganisms that cause various diseases such as diarrhea, hepatitis, cholera and typhoid fever. In addition to microorganisms, river and lake waters contain many particles that also need to be removed before human consumption. There is therefore a need to treat water so that it is fit for human consumption again. When people think of treated water, they usually associate it with the treatment of water that has been polluted, such as sewage, to make it clean again. It is important to make a distinction between water treatment and sewage treatment: water treatment is based on fresh water found in nature, which contains organic residues, dissolved salts, heavy metals, suspended particles and microorganisms. For this reason, the water is taken from the source to the Water Treatment Plant (WTP). Sewage treatment is carried out on residential or industrial sewage so that, after treatment, the water can be reintroduced into the river, minimizing its impact on the environment [Quimica Ambiental, 2010].

The basic activity of filtering/clarifying water for human consumption is among those that generate the greatest volume of waste. The sludge produced in drinking water filtration operations is basically composed of calcium hydroxide, aluminum sulphate, calcite and sand, and generally has a moisture content of over 80% by mass [Raupp-Pereira et al., 2004; Raupp-Pereira et al., 2007].

As *already* mentioned, one sector that has enormous potential to contribute to solving environmental problems arising from a wide range of industrial processes is red ceramics. In fact, studies in the literature have shown that it is possible to recycle industrial waste as a constituent of ceramic masses for the manufacture of ceramic products for civil construction. The use of WTP sludge can be achieved by incorporating it into a red ceramic matrix.

Numerous published studies have evaluated the response of red ceramics when varying percentages of WTP sludge are incorporated into the composition. All of them concluded that this addition alters the mechanical properties of the product, but the product remains within the limits established for civil construction [Teixeira et al., 2002b; Oliveira et al., 2004; Paixao, 2005; Teixeira et al., 2006; Monteiro et al., 2007].

Research into the addition of WTP sludge in concrete matrix carried out by Hoppen et al. [2005] concluded that concrete containing up to 5% sludge can be applied in situations that can range from the manufacture of concrete artifacts, blocks and parts such as covers for manhole covers and manholes, and pedestals for supporting equipment, to the construction of Portland cement concrete sidewalks. For contents above 5%, its use is restricted to applications in which workability is not a key parameter, such as subfloors, sealing blocks and slabs, decorative pieces, sidewalks and residential sidewalks, among others. These authors also found that the low strength of the material containing 10% sludge was probably due to the high consistency, which compromised the densification and quality of the specimens. Although there is a reduction in concrete quality with the

incorporation of WTP sludge, its use as a form of immobilization and co-disposal of waste can be considered interesting from an environmental point of view.

Junior et al. [2006] researched compositions containing 35, 45 and 50% WTP sludge, using 30% of a mixture of glass microspheres used in sandblasting and battery neutralizing salts, completing the rest of the composition with clay. These authors obtained a high mechanical strength value (between 6.6 and 17.1 MPa), which makes it possible to use this waste for the production of ceramics.

2.1.2.4. Residual clay

Brazil is a major producer of ceramic raw materials, due to the large number of clay deposits in the various areas of the ceramics industry. However, the vast majority of these natural reserves are unknown, so there is no technical-scientific data to guide their use and industrial application, as well as their use in a more rational and optimized way by the industrial sector [Menezes et al., 2001].

Brazil has important deposits of industrial minerals for ceramic use, the production of which is mainly concentrated in the southeast and south, where the country's largest ceramic centers are located. The reserves of clay for red ceramics are large and spread over practically all regions of the country [ABCERAM, 2010].

Mining is generally identified as the first stage of the ceramic tile production process. Although the production process generally involves significantly advanced technologies at various stages of production, there is no tradition in this sector of using more modern technologies at the mining stage [Tomi et al., 2000].

Through the specific techniques applied to obtain minerals, both by open-cast and underground methods, regardless of the magnitude of the operation, mining imposes a heavy process of alteration on the exploited area and its surroundings.

Mining a natural resource that is essential to human survival causes changes to the environment that can be minimized. In this case, it is necessary to recover the mined area, which should be understood as reincorporating the area into the local landscape, and not as returning the area to the situation it was in before mining, which would be impossible [Pires, 2000].

Due to the abundance of a wide variety of mineral resources in Brazil, the tradition of mining dates back to colonial times. Today, the Brazilian mining industry is on a par with the world's main mining centers, both in terms of the technologies applied and the quality of the trained professionals and the geology and mining planning research centers [Tomi et al., 2000].

Clay is a natural material with an earthy texture, fine-grained, made up essentially of clay minerals, which may contain other minerals that are not clay minerals (quartz, mica, pyrite, hematite, etc.), organic matter and other impurities. Clay minerals are the characteristic minerals of clays; chemically they are hydrated silicates of aluminum or magnesium, containing in certain types other elements such as iron, potassium, lithium and others. Thanks to clay minerals, clays develop a series of properties in the presence of water, such as: thixotropy and pseudoplasticity in aqueous suspensions, plasticity, wet mechanical strength, linear drying shrinkage and compactability, which explain their wide variety of technological applications [ABCERAM, 2010].

The supply of natural raw materials for ceramics production has some peculiar characteristics. For example, several companies maintain and exploit their own mines, or even control mining companies that supply them with products [Vieira, 2002].

A mining project can be developed over a very variable period of time and involves the following phases: *Design* - the phase of preparing engineering studies and projects, containing construction specifications, necessary for the execution of works; *Implementation* - installation of the project, development of the mine or its preparation for mining; *Operation* - referring to mining activities; and *Decommissioning* - closure of the mining project, with preparation of the area for future use [Cunha, 2003].

Extracting the raw material is the first stage in the ceramics production process. However, only companies with their own clay deposits start at this stage. When extracting the raw material, a first selection must be made, i.e. avoiding layers of clay that contain excess stones and pieces of wood and impurities in general, as these elements could cause obstructions and wear on the molding equipment and compromise the quality of the product [Vieira, 2002].

Modesto et al. [2003] used a clay considered to be waste due to its high iron oxide content to make ceramic tiles. Two other residues from two different effluent treatment plants at a ceramics company were added to this material. These researchers obtained water absorption and mechanical strength results within the norm normally used in the ceramics industry, proving the possibility of producing ceramic flooring using only waste materials.

2.1.3. Classification of Solid Waste

Waste is classified according to the potential risks to public health and the environment, defining its proper handling and final destination.

Considering the need to define minimum procedures for waste management, aiming to preserve public health and the quality of the environment and considering that preventive actions are less costly and minimize damage to public health and the environment, technical standards have been established for this purpose.

Industrial solid waste is all solid or semi-solid waste resulting from industrial, domestic, hospital, commercial, agricultural, service and sweeping activities. Included in this definition are sludges from water treatment systems and those generated in pollution control equipment and installations, including sludges and liquids whose characteristics make it unfeasible to discharge them into the public sewage system or bodies of water or which require technical and economically unfeasible solutions in view of the best available technology.

According to ABNT Standard NBR 10004 of 1987, industrial solid waste is classified into the following classes:

a) Class I waste - Hazardous

Waste which, due to its physical, chemical and infectious properties, may present a risk to public health and the environment. They must have at least one of the following characteristics: flammability, corrosivity, reactivity, toxicity and pathogenicity.

b) Class II Waste - Non-hazardous

- Class II A - Non-inert

Those that do not fall into the classifications of class I or class IIb waste. They have properties such as combustibility, biodegradability or water solubility.

- Class II B - Inert

Any waste that, subjected to static or dynamic contact with water, does not have any of its components solubilized at concentrations higher than the water potability standards defined in Annex H, Table 2.1, of Standard NBR 10004.

Table 2.1: Coding of some waste classified as non-hazardous (Annex H)

Identification Code	**Identification Code**	**Identification Code**	**Identification Code**
A0001	Restaurant waste (food waste)	A009	Wood waste
A004	Ferrous metal scrap	A010	Textile waste
A005	Non-ferrous metal scrap (tin, etc.)	A011	Non-metallic mineral waste
A006	Cardboard waste and paper	A016	Sand casting
A007	Polymerized plastic waste	A024	Sugarcane bagasse
A008	Rubber waste	A099	Other non-hazardous waste

2.2. COMPLIANCE

The forming technique used for a particular application depends on the consistency of the system (e.g. barbotine, plastic paste or granulated material) and will produce a particular shape, with a certain composition and microstructure.

Shaping is a fundamental stage in the manufacture of any ceramic article. Not only because it determines the final geometry and thus its function, but also, and above all, because shaping has to combine the properties of the raw materials in a way that allows the subsequent stages of the production process to be completed successfully. Modeling is correlated with the very concept of ceramics, its history and its technological development. [Handle, 2007].

Extrusion is a forming process widely used in the red ceramics industry to make tiles and bricks. It is a production technique associated with high productivity.

There are several factors that affect the performance of the process and the quality of the final products in extrusion, including: the composition and preparation of the ceramic masses, the plasticity of the masses, the molds (nozzles) used, the type of extruder and the quality of the vacuum [Ribeiro et al., 2003]. This molding technique is mainly used to manufacture ceramic objects with a constant cross-sectional area, and the length of the elements is established by cutting the extruded material perpendicular to the direction of the extrusion flow [Handle, 2007].

The flow conditions of the clay material through the extruder basically depend on its plasticity, which is why it is necessary to keep the characteristics of the mass (composition, degree of grinding, moisture content, etc.) as constant as possible, so that the plasticity value is maintained [Ribeiro et al., 2003]. The consistency of the material used in the extrusion process is completely different from that used in pressing: finished extruded bodies actually have a moisture content of over 1415% and, depending on the raw materials, this figure can be over 2022% [Handle, 2007].

In masses with high moisture contents, close to the liquid limit, there is easy slippage between particles, so the clay mass tends to stick to the extruder's blades and flow through the center of the die at a higher speed. Something similar will happen if the moisture content is reduced and the extruder is working below the zone of maximum plasticity. Friction on the walls of the extruder increases and the clay mass will also tend to flow faster towards the center, while the compression forces developed inside the extruder and the wear of the various metal components increase [Ribeiro et al., 2003].

There are only a few technologies which, in the process of their development, have found such a wide range of applications as extrusion technology. It originated in structural ceramics and even today extrusion is used in industries as diverse as plastics, chemicals, food, etc.

Extrusion has the following advantages over other forming processes:

- High quality and uniform products,
- Versatility: a wide variety of products can be produced by changing a few ingredients and the operating conditions of the extruder,
- Reduced costs: the process has low costs and high productivity compared to other processes,
- High production speed and
- Process automation, reducing manpower.

2.2.1. Plasticity

The term plasticity refers to a particular mode of mechanical behavior whereby a material exhibits permanent deformation without rupture in response to an applied compressive load (limit stress) on the material [Reed, 1995].

Plasticity is a property that is difficult to characterize, since the greater or lesser plasticity of a mass depends on multiple aspects, including: the moisture content of the mass, the different types of clay minerals present (the composition of the mass), the geometry of the crystals and even the particle size distribution of the particles involved. There is also the fact that the determination of plasticity often depends on the skill of the operator and therefore provides a purely qualitative assessment. Furthermore, the values obtained for different tests are not directly comparable [Ribeiro et al., 2003].

The particles that show plasticity are mainly clay minerals. Minerals such as quartz and feldspar do not develop plasticity. Clay is the fraction of the soil that acquires plasticity when it comes into contact with water.

Clay minerals are basically hydrated aluminum silicates, which exhibit plasticity, cation exchange, generally submicrometric dimensions and a lamellar or elongated shape. Clay minerals comprise a large family of minerals, which can be classified into several groups according to their crystalline structure and similar properties. The main groups of clay minerals are kaolinites, illites and montmorillonites [Ring, 1996].

At the beginning of the 20th century, the Swedish chemist Albert Atterberg carried out research into the properties of fine soils (consistency). According to Atterberg, fine soils vary in their state of consistency depending on their moisture content. In other words, soils have different consistency characteristics depending on their moisture content. There are limiting moisture contents that have been defined as consistency limits or Atterberg limits. The term consistency refers primarily to the degree of resistance and plasticity of the soil, which depends on the internal bonds between the soil particles. Cohesive soils have a plastic consistency between certain moisture limits. Below these levels they have a solid consistency and above a liquid consistency. A semi-solid consistency can also be distinguished between the states of plastic and solid consistency [Mitchell, 1976].

The plasticity of a clay soil is related to the shape of its particles, which is characteristic of the clay minerals in the soil. Several authors have tried to correlate consistency limits with the mineralogical aspects of clays. Table 2.2 shows the values of the consistency limits of some clay minerals.

Although the liquidity and plasticity limits can be obtained through fairly simple tests, the physical interpretation and quantitative relationship of their values with the factors of soil composition, type and quantity of minerals, type of adsorbed cation, particle shape and size is difficult and complex [Reed, 1995].

Table 2.2: Consistency limits [Mitchell, 1976].

Clay-minerals	**Liquidity Limit LL (%)**	**Plasticity Limit LP (%)**	**Concentration Limit LC (%)**
Montmorillonite	100-900	50-100	8,5-15
Ilita	60-120	35-60	15-17
Kaolinite	30-110	25-40	25-29

Existing methods for measuring and characterizing plasticity include the Atterberg plasticity index, the Pfefferkorn plasticity index and the plasticity index using the Casagrande apparatus.

2.2.1.1. Liquidity Limit (LL)

The liquidity limit test indirectly measures the shear strength of the soil for a given moisture content, through the number of blows required to slide the slopes of the sample; for a moisture content equal to the liquidity limit, values equal to 2.5 kPa were found, which are very low, indicating the proximity of the liquid state and with most of this resistance being due to the attractive forces between the particles, which in turn are related to the surface activity of the clay minerals [Mitchell 1976].

The liquidity limit of a soil is the moisture content that separates the liquid from the plastic state of consistency and for which the soil has a low shear strength. The test uses the Casagrande apparatus, where both the equipment and the procedure are standardized by ABNT/NBR 6459/84.

2.2.1.2. Plasticity Limit (LP)

The plasticity limit is the minimum moisture content at which the cohesion is small enough to allow deformation, but high enough to guarantee the maintenance of the shape acquired. This limit is the lower end of the range of moisture content at which the soil behaves plastically [Mitchell 1976].

To carry out this test, all that is needed is a glass plate with a ground surface and a standard cylinder with a diameter of 3mm. The test begins by rolling a soil sample with an initial moisture content close to the liquidity limit over the ground face of the plate until two conditions are simultaneously reached: the roller has a diameter equal to that of the standard cylinder and cracks appear (fragmentation begins). The moisture content of the roller under this condition represents the plasticity limit of the soil. The test is standardized by NBR 7180/84.

2.2.1.3. Plasticity Index (PI)

Of the various indices relating the liquidity and plasticity limits and sometimes the moisture content of the soil, the most widely used today is the plasticity index. Physically, it represents the amount of water that would need to be added to a soil for it to go from a plastic state to a liquid state. It is defined as the difference between the liquidity limit and the plasticity limit. This index determines the plasticity of a soil, so the higher the "IP", the more plastic the soil will be. It is also known that clays are more compressible the higher the "IP". According to Caputo [1998], soils can be classified as:

- Weakly plastic 1 < IP 7≤
- Medium plastic 7 < IP≤ 15
- Highly plastic IP > 15

The results of research carried out by Arthur Casagrande led to the creation of a graph, Figure 2.1, which is used to classify a soil according to its plastic properties.

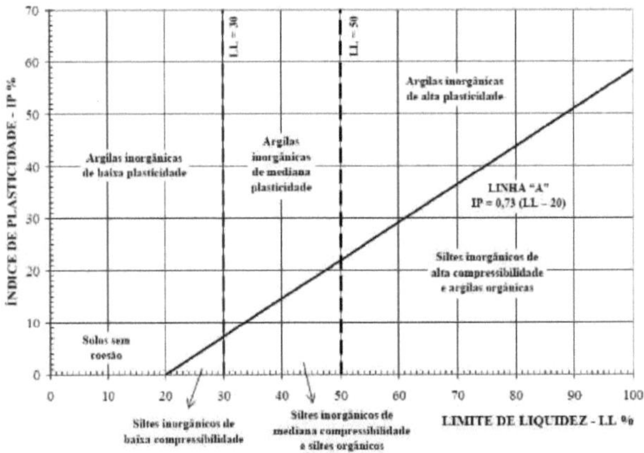

Figure 2.1: Plasticity graph [Holtz and Kovacs, 1981].

2.2.2. Drying

Drying is a key stage in the ceramic production process. The time it takes to lose moisture to the environment cannot be so fast as to damage the product with cracks and warping due to the change in volume. Therefore, in the early stages of the drying process, the clay particles are virtually surrounded and separated from each other by a thin film of water. As drying progresses and the water is removed, the interparticle separation will decrease, manifesting itself in the form of volume contraction. Drying in the internal regions of a specimen is achieved through the diffusion of water molecules to the surface by evaporation. If the evaporation rate is higher than the diffusion rate, the surface will dry more quickly than the interior, with a high probability of defects forming, so the evaporation rate must be lowered for water diffusion to occur. The rate of water evaporation can be controlled by temperature, humidity and air flow rate [Callister, 1999].

2.2.3. Sintering

The purpose of the sintering stage is to agglomerate the particles in order to densify the material and improve its mechanical strength. The sintering stage is where the necessary transformations take place so that the manufactured products achieve the appropriate final properties and microstructure for the type of application designed.

The process parameters; such as temperature, time, pressure, heating or cooling speed and the nature of the atmosphere in which the heat treatment takes place, directly influence the properties of the final product and its microstructure.

During the sintering process, a liquid phase can form, influencing the coalescence between the clay mineral particles and filling part of the pore volume. The degree of vitrification depends on the sintering temperature and time, as well as the mineral composition of the mass that makes up the piece. The fluid phase flows around the unmelted particles that remain in the medium and fills the pores as a result of surface tension forces. In this process, the coalescence of the particles influences the reduction in volume of the ceramic piece. On cooling, the molten phase forms a vitreous matrix which results in a dense and resistant body [Callister, 1999].

2.3. SYSTEM FORMULATION

A ceramic mass must have an adequate chemical and mineralogical composition, so that the physical and chemical transformations that take place during the firing process give the finished product the desired properties. A ceramic mass can be considered suitable when the ratio between plastic and non-plastic raw materials is such that it gives the formed piece sufficient green and dry mechanical strength.

Ceramic materials are inorganic compounds containing at least two, and often more, constituent compounds. In technological applications, they are used in a variety of morphologies including monoliths, powders, thin or thick films, and long or short fibers. They can be mono- or polycrystalline, and can include one or more phases [Chiang et al., 1997].

Phase diagrams are a clear and concise method of graphically representing the equilibrium state for a given composition, temperature and pressure [Kingery et al., 1976]. The use of phase equilibrium diagrams in the study of ceramic materials is becoming increasingly important. After the sintering stage, in some limited areas of the structure new crystalline phases will appear, in others the fusion will have been complete and, in others, the initial raw material may be unchanged. Segadâes [1987] mentions some important information that can be obtained from phase diagrams, such as:

- Melting temperature of each pure component;
- Melting temperature when two or more compounds are mixed;
- Quantities and composition of liquid and solid phases at a specific temperature and mass fraction of the components;
- Interaction of two compounds to form a third compound (e.g. SiO2 and Al2O3 to form mullite, 3Al2O3.2SiO2).

2.3.1 Phase diagram for formulating ceramics from waste materials

Raw materials for the ceramic materials industry can belong to one of three large groups, which will play different roles during processing: plastic components (e.g. clays), fusing components (e.g. feldspars) and inert or dimensional components (e.g. quartz sands). With a view to being used as alternative raw materials, residues can also be classified into three groups: combustible residues, which contain a high content of organic or carbon-rich substances (high calorific value); fluxing residues, which promote better sintering of ceramic bodies (vitreous phase formers); and plasticity-controlling residues, which most often affect the preparation of the green body (dimensional control, both during drying and firing) [Dondi et al., 1997]. Each residue will have a dominant and specific effect on the manufacturing process, but can also contribute to the role played by the other raw materials. Residues that can easily be elevated to the category of alternative raw materials for the ceramics industry are those that are considered inert, safe and produced in large quantities [Segadaes et al., 2005].

In terms of chemical composition, the vast majority of these residues, as well as the natural raw materials (plastics, fluxes or inert materials), contain silica, alumina and lime as their main components. They all contain small amounts of other components, which in most cases will affect the color of the fired product (Fe_2O_3, MnO, TiO_2, Cr_2O_3) but will not play an important role during the processing of ceramics in air at low temperatures. In all cases, the other components in smaller quantities (MgO, K_2O, Na_2O) act as fluxes and can have a strong effect during sintering [Segadaes, 2006].

In order to assess the possibility of a given residue being used directly as a raw material or additive in ceramic processes, or to predict which pre-treatment steps will be necessary for a future use, the selected residue must be characterized. A complete characterization should include an assessment of the chemical and mineralogical composition, particle size distribution and thermal and dilatometric behaviour. For the sake of comparison, a typical clay mixture used normally should also be investigated. Selected combinations of such materials are then used to prepare test samples, which are dried and sintered and also characterized (shrinkage, porosity, density, flexural strength). The selection of the composition and heat treatment is the stage where the use of the phase diagram can be very interesting [Segadaes, 2006].

Most texts on equilibrium diagrams don't go into enough detail to explain the usefulness of phase diagrams for an application, or how to make practical use of the information contained in the diagrams. Often the diagrams were originally prepared using very long times, at one temperature, and nowadays the firing cycle of ceramic products is done in the shortest possible time (in some products the firing cycle is less than 30 minutes from cold to cold), and are therefore very far removed from the conditions used to determine the phase diagram.

Although under normal industrial operating conditions thermodynamic equilibrium is usually not reached, the phase equilibrium diagram of the relevant system can still be used to predict the tendency of the reaction to complete and will be a great help while making appropriate choices of compositions and process parameters [Raupp-Pereira et al., 2006]. Most ceramic engineers may not need to know how to create a phase diagram, or how to calculate one from the relevant thermodynamic data, but it can be very useful to know how to read, interpret and apply the relevant diagram information to a particular process [Segadaes, 2006].

In the ceramics industry, when natural raw materials are replaced by industrial waste, if it is necessary to predict the effect and/or role of the waste, this can be observed within a relevant system of a phase diagram. The first difficulty is choosing a relevant system, which would be at least SiO_2-Al_2O_3-CaO- MgO-Na_2O-K_2O-Fe_2O_3. However, such diagrams of systems containing multiple components are rarely available. Therefore, a restricted diagram, preferably of a ternary system, has to be selected, and it will not necessarily be defined only by the major components of the mixture. Once the relevant system has been shrunk, the second difficulty arises: calculating the equivalent composition of the mixtures, because the other components should not simply be ignored. To be able to make an appropriate decision, a good knowledge of the science and processing of ceramics is necessary [Segadaes, 2006].

The phase diagram of the relevant system can provide valuable estimates of the proportions of phases present during and after sintering, as well as the presence of liquid phase at sintering temperature and the resulting crystalline phases [Raupp-Pereira et al., 2004].

The melting effect of K_2O and Na_2O is quite similar, and their phase diagrams with alumina and silica are morphologically similar. K_2O leads to the formation of a liquid phase at lower temperatures and with lower viscosity, but the effect of Na_2O is even greater. Therefore, it will be better to discuss the effect of alkalis in terms of the phase equilibrium diagram for the Al_2O_3-SiO_2-$K_{(2)}O$ system, using the K_2O+Na_2O junction as the K_2O equivalent. For alkaline earth oxides (CaO+MgO), their contribution can be ignored or added to the alkaline oxides. In other words, the effect of melting oxides on the Al_2O_3-SiO_2 system can be discussed in terms of the combination of the K_2O+$Na_{(2)}O$

components, or better still, the combination of $CaO+MgO+K_2O+Na(2)O$ [Segadaes, 2006].

3. MATERIALS AND METHODS

3.1. MATERIALS

In a preliminary analysis, mineral residues from the state of Santa Catarina were selected which could potentially be used as alternative raw materials in the ceramics industry. Only residues from the state of Santa Catarina were analyzed due to the transportation costs involved, but the selected residues are easily found in other regions.

In the municipality of Araquari (Pedras Morro Grande company), a slurry is produced from the process of crushing gneiss rock. Sludge from the cutting of ornamental varvito rock, from the municipality of Trombudo Central (Ramos Universo company), was also one of the residues used in this study. Another waste used in this study, from the water treatment plant, was WTP sludge (SAMAE, Blumenau). The Eliane company, located in the municipality of Cocal do Sul, provided a clay which is considered to be waste due to its high iron content. The municipalities where the companies supplying the waste are located are shown in Figure 3.1.

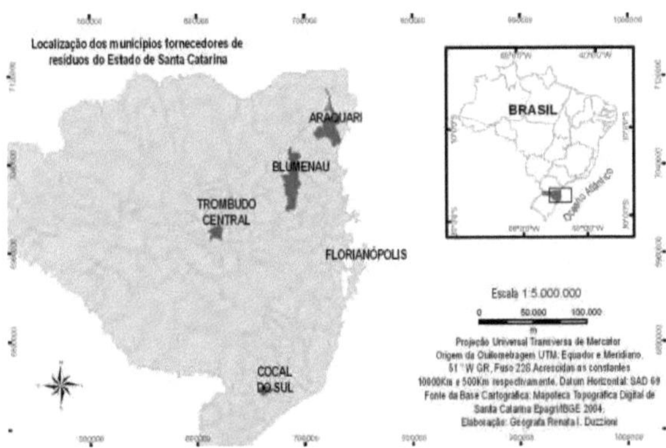

Figure 3.1: Municipalities where the companies supplying the waste are located.

3.1.1. Gneiss

At the company supplying this waste (Pedras Morro Grande, Araquari, SC), gneiss slurry is produced during the gneiss rock crushing process. This process includes breaking and screening the rocks until they reach market specifications. This circuit can generate crushed stone ranging in size from pebbles to boulders.

Crushing is the primary comminution process. The number of crushing stages depends on the size of the material in the feed and the quality of the final product. The particle size and morphology of the gravel is controlled throughout the secondary, tertiary and quaternary stages (sand making). The waste generated directly from the rock comminution process is directed to a water wheel. The purpose of this wheel is to separate the crushed stone, which can still be sold, from the fines from the process, which are considered waste.

The suspended fines are separated from the material that can still be used by means of a metal plate with several 5 mm openings.

The finest raw material passes through the water wheel, which uses gravity to separate the raw material from the suspended fines. The material retained in the water wheel has commercial value and is sold by the company as gravel for the construction industry. The fines from the process, which have no commercial value, are removed from the water wheel through a pipe and deposited in settling ponds at the company's own plant, as shown in Figure 3.2.

Figure 3.2: Settling pond for gneiss waste.

The production of this waste reaches 700 m^3/month. When the maximum volume of the settling pond is reached, it is cleaned, i.e. the residue is removed from the pond by a backhoe and is usually accumulated on the company's own land or sent to industrial landfills. The gneiss waste used in this work comes from the first settling pond.

3.1.2. Varvito

In Trombudo Central, the region from which the waste under study comes, varvito has been mined since the 1960s in the town of Bracatinga I, where there are several deposits. Mining began in a rudimentary way, using metal levers and wedges, aided by dynamite explosions. The rocks extracted were used without any processing and had a fairly regular symmetry [Cunha, 2007]. The commercial name for this ore in the region is slate (metamorphic rock). However, the rock found in this region is in fact of sedimentary origin, rhythmic shale, known as varvite. One of the varvito quarries found in Trombudo Central is shown in Figure 3.3.

Figure 3.3: Trombudo Central quarry.

In varvite mining, the waste can be called sludge, dust or fines from cutting and polishing the pieces, and the slabs that break off during cutting or even after polishing.

Much of the region's physical needs are met by the process of mining and extracting varvito, such as paving, building walls, roofing, constructing bus stops and decorative structures related to entertainment.

The company Ramos Universo, the supplier of varvito waste, has its production entirely dedicated to the export of slabs used in the production of billiard table tops. The process begins with the removal of the varvito blocks from the quarries (Figure 3.4). The blocks, measuring 1.38× 0.88 or 1.58× 1 m, are cut and removed from the quarry by tractors, loaded onto trucks and taken to the company.

At the company, the rock processing process involves 8 stages. The first stage involves splitting the blocks into smaller slabs. The splitting can be done manually, when the rock has veins that make this work possible, or by using a deburring machine, as shown in Figure 3.5.

Pre-calibration, the second stage, is responsible for the most waste production, as it consists of wearing down the plates by rubbing them with a diamond ring (Figure 3.6). The company works with two slab thicknesses: 24 mm and 30 mm. Most of the cutting stages involve the use of water, generating a flow of water and residue that goes down an internal pipe straight into the first settling pond.

Figure 3.4: Removing the blocks from the quarry.

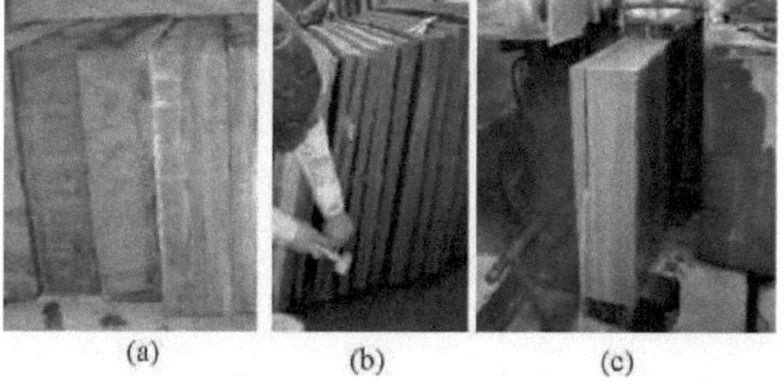

Figure 3.5: Block splitting (a) blocks, (b) manual splitting and (c) mechanical splitting.

After pre-calibration, the boards move on to the cutting stage, where the boards are *already* cut to final length. The fourth stage is the drilling stage, both for the pockets (6 holes) and the fixing points (30 holes). After drilling, the three-piece sets that make up a billiard table go for final calibration (fifth stage). All the waste generated during the process is taken through internal pipes to the decantation ponds.

Figure 3.6: Pre-calibration stage.

There are six settling ponds, Figure 3.7. The largest amount of material is retained in the first lagoon, Figure 3.8. The waste used in this work comes from this first lagoon. The other settling ponds have the function of further settling the material before the water is returned to the river. In the last two ponds, the water quality is good enough to allow the owner to raise 1,200 fish.

Figure 3.7: Settling ponds for varvito waste.

As most of the processes involve the use of water, the sixth stage consists of leaving the boards for 72 hours in an oven at approximately 40°C. After the period in the oven, the boards go to the curing room, where they remain for 30 days to acclimatize (humidity and temperature) so as not to cause warping in the future. The final stage consists of gluing the MDF boards together and curing them for 24 hours. Figure 3.9 shows the billiard table in its final stage.

Figure 3.8: First settling pond, provider of the waste used in this research.

Figure 3.9: Billiard table board ready for export.

The first settling pond is approximately 15m long and 8m wide, with a depth of 1.5m. The waste is removed from the lagoon by a backhoe and placed on the edge of the lagoon to remove excess water. The lagoon is cleaned every 10 days and produces around 250 m^3 of waste per month. The Ramos Universo company has two plots of land with environmental licenses for the disposal of this waste, but has shown great interest in giving this waste an environmentally correct destination.

3.1.3. Water Treatment Plant Sludge

The sludge resulting from the cleaning of drinking water was also one of the residues used in this work. At the Autonomous Municipal Water and Sewage Service (SAMAE) in Blumenau, the water purification process begins with the addition of aluminum sulphate when the raw water enters the treatment plant. The introduction of the aluminum sulfate is intended to agglomerate the solid particles on the walls of the tanks and also on the containment plates, to facilitate future removal. In the next stage of the process, the water passes through a chute and enters the flocculator, where it is stirred slowly so that the particles increase in size, forming flocs.

The next stage is decanting, in which the impurities that have agglutinated and formed flocs are separated from the water by the action of gravity, going to the bottom of the tanks or getting stuck on their walls. The next stage is filtration, where the water passes through large filters with layers of pebbles (river stones) and sand, with different granulations and anthracite coal (mineral coal). The filter retains the impurities that have passed through the previous stages. The water at this point is already potable, but for greater protection against the risk of water-borne infections, a disinfection process is carried out.

Disinfection, also known as chlorination, is used to eliminate germs that are harmful to health and guarantee the quality of water all the way to the consumer's tap. In this process, sodium hypochlorite, chlorine gas or chlorine dioxide can be used.

The next step is fluoridation, where sodium fluorsilicate or fluorsilicic acid is added in appropriate dosages. The function of fluorsilicate is to prevent and reduce the incidence of dental caries, especially in consumers between the ages of zero and 14, the period when teeth are formed.

The last action in this water treatment process is pH correction, when hydrated lime or light briquettes (sodium carbonate) are added to ensure adequate neutralization to protect the mains pipes and users' homes. Around 30 minutes elapse between the time the raw water enters the WTP and the time it leaves, ready for drinking, as shown in the diagram in Figure 3.10.

During the decanting process, water treatment plant sludge (WTP sludge) is generated by the impurities retained in the natural water potabilization process and the chemical products used in the treatment. SAMAE is currently looking for a viable alternative for using this waste, so that it is no longer sent to industrial landfills, generating costs for the company. This sludge contains organic matter from the raw water and impurities contained in the chemicals used in the treatment, such as: aluminum, iron, manganese and chromium, among others.

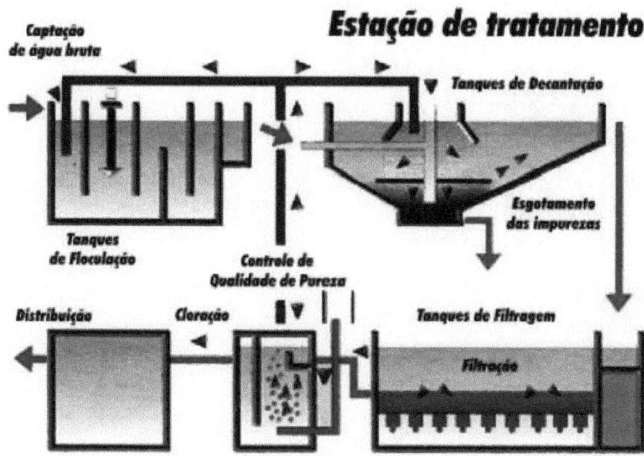

Figure 3.10: SAMAE's water purification process [SAMAE, 2008].

Plants and tanks are cleaned once a week and the amount of sludge removed depends on the turbidity of the water. On average, two gallons of 300 l/week are removed, which generates around 2,400 l/month (with a water percentage of around 63%).

3.1.4. Residual clay

The fourth waste selected for this study is a clay considered to be waste by the company that owns the deposit (Eliane, Cocal do Sul, SC) due to its reddish firing. Yellowish and reddish clays can occur both above the light clay deposit and below and inside the light clay layer, in the form of intercalations or contaminated pockets. When present at the same stratigraphic level as the light clays, the yellowish and reddish clays are discarded when selecting the clays used in ceramic processes. During the mining process, the hydraulic excavator selects the light clays in the operation to dismantle the clay layers of the deposit, throwing the yellow clays into the mined pit as backfill.

The waste used in this work, from the extraction of clay for the ceramics industry, is part of the upper layer of the deposit. This layer of clay contains impurities that make it undesirable in the manufacturing process. Because it contains a large amount of iron oxide (Fe_2O_3) and because it has a reddish color after heat treatment, which would darken the base of the ceramic tiles, this clay is treated as waste. The residual clay used in this work is part of the group known as "third clay", Figure 3.11.

Figure 3.11: Residual clay stock, yellowish in color.

The production of yellow clay and the formation of stockpiles will only occur at the request of potteries, for example, or as a remnant of the exploitation of an exhausted deposit, when higher iron contents are allowed in third-party clays.

The Eliane company has its own quarry and the residual clay taken from the upper layers, as mentioned above, is only stored and not used in any of the company's putty formulations. The stock of yellow clay is currently around 3,000 tons, which can be increased on request.

3.2. EXPERIMENTAL METHODOLOGY

The experimental procedures related to this work were carried out at the Federal University of Santa Catarina, Florianópolis, SC, at the Materials Laboratory (LABMAT) and Microstructural Characterization Laboratory (LCM) of the Mechanical Engineering Department and at the Materials Technology Center (CTCMat), located in Criciùma, SC, as well as at the University of Aveiro (Portugal), at the Ceramic and Glass Engineering Department (DECV).

The flowchart shown in Figure 3.12 briefly illustrates the methodology used to choose the formulations used to make ceramic tiles. Each item will be covered in detail.

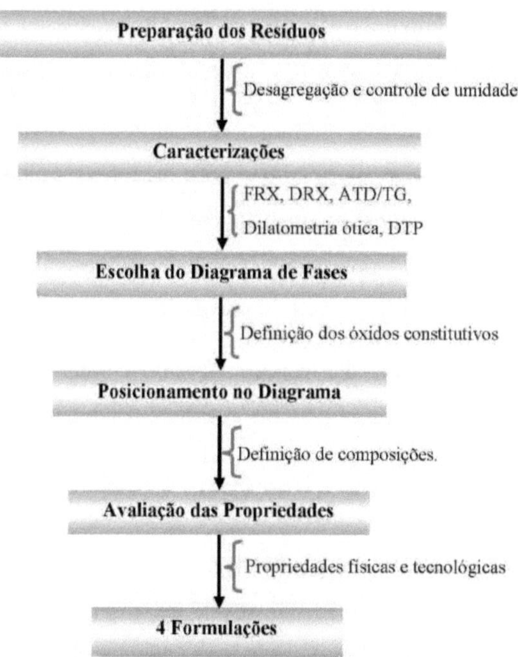

Figure 3.12: Flowchart of the methodology used to prepare the formulations.

As most of the waste is sludge, it required prior preparation. The waste from crushing and polishing the gneissic and varvitic rocks was dried in an oven at 110°C until it was a constant mass, and then disintegrated in a ball mill for 5 min. The residual clay was only subjected to a reduction in humidity to facilitate its disintegration. The WTP sludge, which contains more than 50% water, required filter-pressing. This operation was carried out at the T-Cota company in Tijucas, SC.

The second stage focused on characterizing the waste used, with the aim of investigating the potential and technological limitations of using it as a raw material in the production of ceramic tiles.

Through chemical analysis by X-ray fluorescence, the percentages of each oxide constituting the residue were determined. The phases present were identified using X-ray diffractometry and, in order to verify their behaviour during burning, as well as possible phase transformations and mass losses, the residues were characterized using differential thermal analysis and thermogravimetry. Laser diffraction was used to determine the particle size distribution. Thermal expansion curves were obtained using optical dilatometry.

With the results of the chemical analysis, it was possible to choose the relevant phase diagram for this work. For this choice, the major components were taken into account, such as the oxides of silicon, aluminum and potassium. The latter represents the combination of alkaline and alkaline-earth oxides (CaO+MgO+K2O+Na2O).

Using softer excel, the chemical compositions of the residues were positioned within the chosen phase diagram. In the field of work limited by the chemical composition of each residue *in natura*, four formulations were selected in order to cover the largest possible area of work.

With the chemical compositions positioned on the phase diagram, it is possible to predict the properties of the ceramic to be obtained. However, the phase diagram represents thermodynamic equilibrium, and in industrial processes, this equilibrium is usually not reached. However, the diagram can still be used to predict the properties of the manufactured material.

After preparing the formulations, the forming method adopted was extrusion. However, press forming was used for some preliminary tests, as it is a simple and practical method to use on a laboratory scale, requiring little material. The extrusion method was used for the final tests in the manufacture of all-waste ceramic tiles, as it is a low-cost and high-production method.

3.3. CHARACTERIZATION OF THE WASTE

3.3.1. Chemical Analysis (FRX)

Quantitative chemical analysis of the four residues was carried out by X-ray fluorescence (XRF) on a Philips PW 2400 machine with a 3 kW tube and rhodium target at CTCMat in Criciùma, SC.

All the samples were homogenized with lithium tetraborate, which acted as a flux, and with lithium nitrate, which was used to reduce possible metals that would cause oxidation in the platinum crucible. These samples were melted at 1000°C in the form of 40 mm diameter pearls, and at the end of this preparation procedure, they were taken to the equipment for analysis.

3.3.2. Mineralogical Analysis (XRD)

X-ray diffractometry (XRD) allows the identification of materials in terms of their crystalline characteristics and the quantification of constituent elements, provided that they are in an appropriate concentration range. This test is applicable to inorganic substances, especially minerals. This technique was used in this work with the aim of identifying the mineralogical phases present in the raw materials used, as well as characterizing the ceramic materials developed in terms of the quantity of crystalline phases present.

Samples of the waste at room temperature were analyzed at the LCM/UFSC. The equipment used was a Philips Xpert diffractometer, with copper Kα radiation (l = 1.5418 Â), nickel filter in the secondary optics, power of 40 kV and 30 mA and 1° divergence slit. The analysis of the residues sintered at 1100 and 1150°C was carried out at the University of Aveiro. The powdered samples were lightly compacted on a glass slide and analyzed in a Rigaku Geigerflex D/max - C series diffractometer, using copper Ka radiation (wavelength $\lambda = 1.541 \times 10^{-10}$ m), 40 kV / 30 mA, with a scanning speed between 4 and 80° 2θ of 6°/min.

3.3.3. Thermal Analysis (ATD and TG)

The thermal behaviour of the raw materials was characterized using the thermoanalytical techniques of differential thermal analysis (DTA) and thermogravimetric analysis (TG).

These analyses were carried out at the University of Aveiro, Portugal, on a Netzsch 409 EP device using a Pt-PtRh thermocouple. The crucibles used were made of alumina and the analyses were carried out in an air atmosphere at a heating rate of 10 °C/min.

3.3.4. Optical dilatometry

The thermal expansion curve of the formulations was obtained using a Misura optical dilatometer, model 3.32, with the collaboration of Expert System Solutions (Modena, Italy). This instrument can provide thermal expansion data like a dilatometer, but has the added benefit of using optical techniques as opposed to mechanical measurement. The device uses a 50 mm long sample that can be pressed from powder or cut from the material to be analyzed. The sample is held in place by refractory rods and can be supported by an alumina plate. The sample is placed horizontally on the alumina plate or directly on the two rods.

The residues were heated to 1300°C with a heating rate of 5°C/min. This test shows the behavior of the materials with increasing temperature, but without the influence of plateaus. With the data obtained from this curve, together with the information obtained from the phase diagram, the firing temperatures of the samples in this work were determined.

3.3.5. Particle Size Distribution

The laser diffraction technique was used to determine the particle size distribution (PSD). A Cilas laser diffractometer, model 1064L, with an analytical range of 0.04 to 500 μm, 64 detectors, 100 particle size classifications, and two laser emitters for better precision in the submicrometer range, was used for this technique. The samples were prepared as a suspension (sample+ water), with 10% solids content by mass and subjected to stirring to completely deagglomerate the particles. The suspension is then inserted into the equipment and the DTP reading is taken.

3.4. CERAMIC FORMULATIONS AND PROCESSING

In order to predict the effect that replacing natural raw materials with waste can have, a relevant ternary system was chosen to guide the preparation of the formulations.

In terms of chemical composition, the dominant oxide in the vast majority of waste is silica (SiO_2), followed by alumina (Al_2O_3) and iron oxide (Fe_2O_3), as well as flux oxides (K_2O and Na_2O). Therefore, based on the chemical composition of the waste, the system chosen was SiO_2-Al_2O_3-K_2O. In this case, the potassium oxide content is considered to be the total of alkaline oxides (K_2O+Na_2O+MgO+CaO), as described in the literature review. This system is shown in Figure 3.13 and was used to formulate suitable compositions for ceramic products.

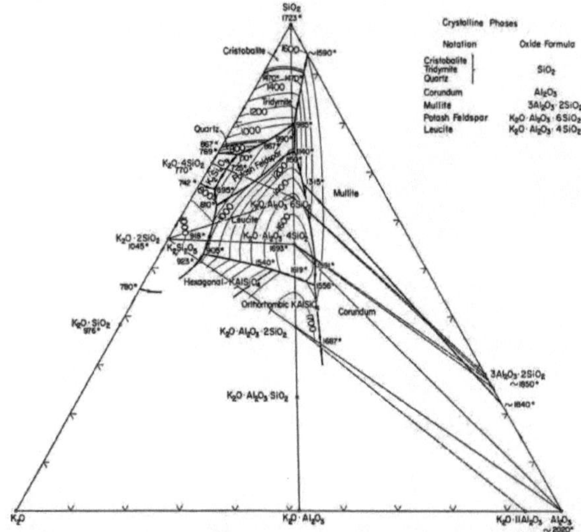

Figure 3.13: SiO2-Al2O3-K2O phase diagram [Levin, 1964].

The formulations were prepared with the waste as received, according to the pre-established percentage. To homogenize the formulations, they were processed three times in a mixer (Gelenski, model LIG-05). Mixing the components is the most critical stage in the process of preparing a paste for extrusion, as insufficient mixing of the various components leads to difficulties in all subsequent stages such as extrusion, handling, drying, sintering and finishing.

The experimental part of this work was carried out at the University of Aveiro, Portugal, where the formulations were packaged and transported to the university.

Due to the small quantity of raw material (transported to Portugal), we opted to use a manual extruder, which is also easy to handle and has a mold for making cylindrical specimens with a diameter of 10mm.

The laboratory extruder used in this work is shown in Figure 3.14. The machine consists of (1) a piston compression system, which compresses the ceramic mass and forces it through the extrusion mouth and (2) the extrusion mouth, which is the opening of the mold that forms the extruded material.

It is not possible to work continuously with this type of extruder. It also has the disadvantage of not homogenizing the dough, which can increase the likelihood of defects.

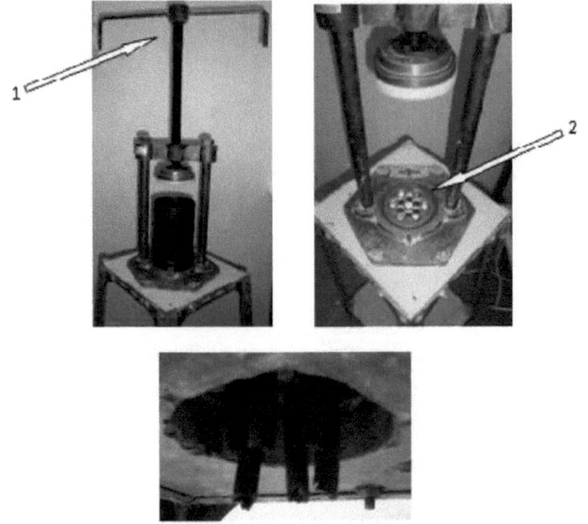

Figure 3.14: Manual extruder.

After extrusion, the pieces were air-dried and sintered at 1100 and 1150°C, with a heating rate of 5°C/min and a dwell time of 40 minutes. A Termolab laboratory furnace with a maximum working temperature of 1700 °C was used to sinter the materials.

3.5. CHARACTERIZATION OF FORMULATIONS

3.5.1. Preliminary Tests

Preliminary tests were carried out on pressed tablets in order to check whether the working temperatures chosen using the phase diagram are really the most suitable. For this purpose, water absorption and density tests were carried out on samples sintered at 900, 950, 1000, 1050, 1100 and 1150°C. Five specimens were made for each temperature by uniaxial pressing at a pressure of 50 MPa. The formulations are composed of four residues with different particle sizes, so a high pressure of 50 MPa was required to aid packing during pressing. The samples were sintered for 40 min. at a heating rate of 5°C/min.

3.5.1.1. Water absorption

The water absorption of a sintered ceramic material is a parameter used to measure open porosity and assess the material's castability. It is defined as the gain in mass, expressed as a percentage, that the pieces show when introduced to boiling water for a given period.

Table 3.1 shows the classification of ceramic tiles according to standard NBR 13818 (1997) in relation to water absorption groups.

The water absorption tests on the sintered specimens were carried out in accordance with standard NBR 13818, with immersion in boiling water for 2 hours.

Table 3.1: Classification and definition of ceramic tiles in relation to water absorption groups.

Groups*		Water absorption (AA, %)	Product definition
AI	BIa	0 < AA≤ 0.5	Porcelain tiles
AI	BIb	0.5 < AA 3≤	Sandstone
AIIa	BIIa	3 < AA 6≤	Semi-gross
AIIb	BIIb	6 < AA≤ 10	Semi-porous
AIII	BIII	AA > 10	Porous

* A = extruded; B = pressed.

To carry out this test, the specimens had to be left for at least 2 hours in an oven and then cooled in a desiccator. The samples were weighed, identified and placed in boiling water. This condition was maintained for 2 hours. The specimens were removed from the boiling water and transferred to a cooling vat for 1 hour. After this procedure, an absorbent cloth was moistened and spread out on a flat surface to remove the surface water from each side of the specimen. The samples were pressed onto the cloth and then weighed.

The following calculation was used to calculate water absorption:

$$AA = \frac{M - M_0}{M_0} \times 100 \qquad \text{(eq. 3.1)}$$

where:

AA = water absorption (%)

M0 = dry mass (g)

M = wet mass (g).

3.5.1.2. Density

The tablets used in the preliminary tests were analyzed for four densities: geometric, *bulk*, apparent and pycnometric. Geometric density is measured from the dimensions and mass of the specimens. For this procedure, the specimens were placed in an oven at 60°C for 2 hours and then in a desiccator for 30 minutes and measured using a caliper.

To calculate the geometric density, the following equation was applied:

$$\rho geo = \frac{m}{V} \qquad \text{(eq. 3.2)}$$

where:

ρgeo = geometric density (g/cm^3) m = sample mass (g).

V= Sample volume (cm^3)

The definition of *bulk* density includes real situations in which samples are composed of pores, cracks, crystalline defects, amorphous phases, etc. *Bulk* density is simply the ratio between the mass of the sample and its total external volume. Bulk density was calculated from the values of open porosity and water absorption.

To determine the apparent density using the Archimedes method, the sample was weighed; then the sample was placed in a sample holder and immersed in water, where the weight of the suspended and immersed sample is measured. The density is calculated using the expression:

$$\rho = \frac{\rho_{ag}.M_A}{M_A - M_I} \qquad \text{(eq. 3.3)}$$

where: ρ_{ag} = density of water (g/cm3) MA = mass of dry sample (g) MI = mass of immersed sample (g)

This technique makes it possible to assess open porosity, but not closed porosity.

In the pycnometric density technique, Quantachrome pycnometer (model Nova 1000), a chamber of known volume was filled at a certain temperature with a certain mass of helium gas, resulting in a certain pressure.

The formulations were characterized by analyses such as X-ray diffractometry, thermal analysis, optical dilatometry, plasticity, linear shrinkage, water absorption and mechanical flexural strength, in addition to the preliminary tests explained in the subsequent sections.

3.5.2. Plasticity

The Casagrande method was used to calculate the plasticity index of the formulations. In this method, the moisture content is determined by closing a groove made in the sample arranged in a metal shell, by means of 25 blows at constant speed of this shell against the fixed base, according to NBR 6457.

To determine the liquid limit (LL), the procedure was divided into stages. The first stage of this test consists of preparing 400g of #80 mesh sample. This sample is then hydrated with a sufficient amount of water so that it does not stick to the hands and then the sample is left to rest for at least 8 hours.

The second stage consists of placing around 200g of the hydrated sample in a porcelain dish, correcting the height of the Casagrande spoon drop with the height check gauge.

Next (the third stage), the sample should be homogenized with the help of a spatula to obtain a homogeneous paste with a plastic consistency, and part of the sample should be transferred to the Casagrande spoon, shaping it so that it fills the entire spoon and the height of the central part is adjusted according to the chisel. Divide the mass in the center of the spoon into two parts, passing the opposite part of the template to check the height of the spoon. The chisel should be placed perpendicular to the surface of the spoon as shown in Figure 3.15.

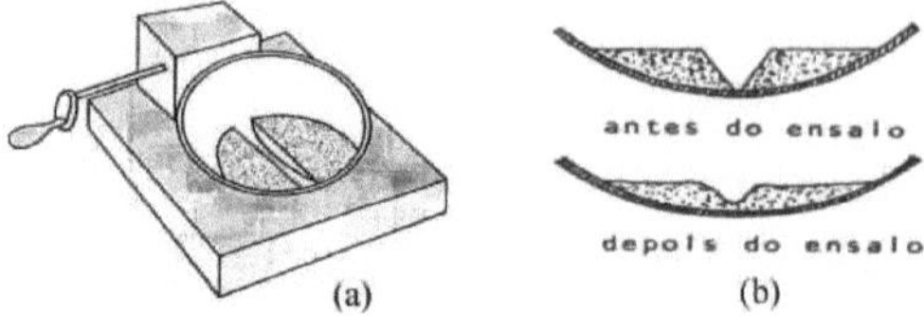

Figure 3.15: (a) Casagrande spoon (b) appearance of the sample before and after testing.

To carry out the fourth stage, the device is held in the base and the spoon is struck at a speed of approximately two revolutions per second against the base by turning the crank on the side, letting the spoon fall freely until the inner edges of the groove join over a length of 13 mm after completing (50± 3) strokes.

A portion is then removed from the place where the sample joined and the moisture content is determined, thus obtaining a pair of values, "moisture content x number of blows", which will define a point on the creep graph. Repeating this procedure for different moisture contents (40, 30, 20 and 10± 3 blows) allows the graph to be constructed. It has been agreed that in the Casagrande test, the moisture content corresponding to 25 blows needed to close the groove is the liquidity limit.

To calculate the liquidity limit, the results obtained from the five points (50, 40, 30, 20 and 10 blows) are taken and the percentage of moisture for each point is calculated as follows:

$$LL = \frac{MU - MS}{MU} \times 100 \qquad \text{(eq. 3.4)}$$

where:

LL = liquidity limit (%) MU = wet mass (g) MS = dry mass (g).

The plasticity limit (LP) is determined by the percentage of moisture at which a sample begins to fracture when you try to shape it into a cylinder 3 mm in diameter and 10 cm long. To determine the LP, about 10 g of the sample must be rolled on the ground glass plate using the palm of the hand so that it has a cylindrical shape similar to that of the template for comparison until it breaks. Then transfer the broken pieces of specimen to a coded, weighed container and record the mass (and repeat this step two more times).

After this procedure, the sample is placed in the oven and left for at least 2 hours. However, if the specimen fragments before reaching a diameter of 3 mm and a length of 10 cm, it should be returned to the porcelain dish and repeated. After removing the containers from the oven, they must remain in a desiccator to cool to room temperature, so that the values can be recorded and the calculations carried out.

To calculate the LP, the results obtained with the three rollers should be taken and the percentage of moisture for each roller calculated as follows:

$$LP = \frac{MU - MS}{MU} \times 100 \qquad \text{(eq. 3.5)}$$

where:

LP = plasticity limit (%)

MU = wet mass (g)

DM = dry mass (g).

The plasticity index (PI) of a material is defined as the difference between the liquidity limit and the plasticity limit of the sample. The following equation is used to calculate the PI:

IP (%) = LL - LP (eq. 3.6)

The sample is characterized as "non-plastic" when it has a minimum moisture content, and even then a point lower than that of 50 strokes is obtained. Thus, it is not possible to remove more moisture to obtain the highest point value; and when even the 50-stroke point is removed, it is not possible to mold the roller at the minimum moisture point, which is 50 strokes.

The results obtained by the Casagrande apparatus can be analyzed according to the criteria used by Caputo [1998]: weakly plastic (1 < IP < 7), moderately plastic (7 < IP < 15) and highly plastic (15 < IP).

3.5.3. Mineralogical Analysis (XRD)

The mineralogical analysis of the formulations sintered at 1100 and 1150°C was carried out at the LCM/UFSC. The equipment used was a Philips Xpert. diffractometer, with copper Kα radiation (l = 1.5418 Â), nickel filter in the secondary optics, power of 40 kV and 30 mA and 1° divergence slit. A sample rotator was used for all readings, with a rotation of 1 rpm, in order to minimize preferential orientation effects. For the characterization of the materials, the powdered samples, with a particle size of less than 45 μm, were compacted in a sample holder for the inverse surface reading. The analysis conditions were: 0.02° step, 2 s step time and measurement range, in 2θ, from 10 to 90°. The ICSD and JCPDS databases were used to identify the phases present.

For the analyses carried out on the formulations after sintering, a piece of each sample had to be ground in a mortar until it reached particle size fractions of less than 44 μm. The analysis conditions were: 0.02° step, step time of 2s and measurement range, in 2θ, from 10 to 90°. The JCPDS database was used to identify the phases present.

3.5.4. Specific Surface Area Analysis

The specific surface area of the formulations was evaluated using the BET method at the Ceramics and Glass Department of the University of Aveiro. The method devised by Brunauer, Emmett and Teller in 1938, better known as BET [Gregg and Sing, 1982], allows the specific surface area of a sample to be determined using the technique of isothermal adsorption of a gas, when a flow of a mixture of adsorbable gas plus non-adsorbable (carrier) gas passes over the sample.

To determine the specific surface area, a Micromeritics - Gemini apparatus was used, which uses the multipoint method (5-point measurement), with degassing of the sample at 200°C for 2 hours and cooling to room temperature. The method used is that of adsorption isotherms, using nitrogen as the adsorbed gas and helium as the carrier gas.

3.5.5. Thermal Analysis

See item 3.3.3.

3.5.6. Optical dilatometry

See item 3.3.4.

3.5.7. Linear Firing Retraction

Linear shrinkage occurs due to a densification of the material as a result of the coalescence of the particles and the formation of a liquid phase during sintering. The liquid phase, formed during sintering, decreases in viscosity as the temperature rises, filling the voids between the particles, allowing for a reduction in porosity and greater densification of the piece.

The linear shrinkage (LR) test is determined based on the change in length of the dry specimens and after they have undergone the sintering cycle. LR measurements were taken using a caliper with a resolution of 0.01 mm. The equation was used to calculate the RL:

$$RL = \frac{L_0 - L_1}{l_0} \times 100 \qquad \text{(eq. 3.7)}$$

where:

RL = linear shrinkage (%)

L_o = length of dry part (mm)

L1 = length of burnt part (mm)

This test shows the behavior of the materials as the temperature increases, taking into account the threshold used, allowing time for the transformations to take place.

3.5.8. Water absorption

See item 3.5.1.1.

3.5.9. Mechanical Flexural Strength

The mechanical strength of a material determines its limitations for applications where the material can withstand mechanical stress [Acchar, 2000]. In the case of traditional ceramic materials, mechanical strength is usually measured by bending. According to Padilha [1997], the presence of porosity and vitreous phase in the structure of traditional ceramic materials has the effect of reducing mechanical strength. The flexural strength test was carried out on a Shimadzu AG-25TA universal tester at a speed of 0.5 mm/min at the University of Aveiro.

3.5.10. Analysis of Fracture Surface Microstructures

The microstructural characterization of the fracture surfaces of the formulations was carried out using scanning electron microscopy (SEM) at the Department of Ceramics and Glass of the University of Aveiro, DECV. A Hitachi SU-70 microscope was used, at 15 kV, after carbon deposition, and the qualitative analysis of the phases observed was carried out by EDS Bruker, Quantax 400.

4. PRESENTATION AND DISCUSSION OF RESULTS

This chapter discusses the analysis of the results of the characterization of the residues and their formulations. A preliminary study on the economic viability of the products resulting from this research has been carried out and is included at the end of this chapter.

4.1. CHARACTERIZATION OF THE WASTE

4.1.1. Chemical Analysis (FRX)

The chemical compositions, in oxides, of the four *in natura* residues investigated *in* this work are shown in Table 4.1. From the XRF analysis, it can be seen that the residues consist predominantly of silica and alumina. Varvito waste has the highest percentage of silica (74.32%) and WTP sludge has the highest amount of Al2O3 (22.10%), along with other oxides in smaller quantities.

The waste with the highest percentage of loss on ignition, related in this case to the constituent water and/or organic matter present, was the WTP sludge with 12.76%. The highest percentage of flux (Na_2O and K_2O) was found in the residue from crushing the gneiss rock (8.79%). The highest amount of iron oxide is found in the WTP sludge and residual clay, with 7.02% and 6.51% respectively.

Table 4.1: Chemical composition, in oxides, of the waste (by mass).

Oxides	Gneiss	Varvito	WTP sludge	Residual clay
SiO2	59,22	74,32	53,30	63,01
Al2O3	16,75	8,79	22,10	19,55
Na2O	4,48	3,12	0,24	0,05
K2O	4,31	1,48	2,11	2,83
CaO	5,98	2,68	0,12	0,04
MgO	1,63	1,74	1,20	0,77
Fe2O3	4,56	2,43	7,02	6,51
MnO	0,14	0,16	0,09	0,02
TiO2	0,43	0,51	0,83	0,91
P2O5	0,75	0,18	0,24	0,07

Fire loss	1,74	4,59	12,76	6,30

According to the oxides found in the chemical analysis of the residues, a system was chosen to help predict the phases present as well as assisting in the choice of process parameters. This system was SiO2-Al2O3-K2O, Figure 4.1, and the effect of melting oxides in the Al_2O_3-SiO_2 system was discussed in terms of the combination of the components CaO+MgO+K2O+Na2O.

A point corresponding to each raw residue was marked on the phase diagram. It can be seen that any formulation made with these four residues will necessarily belong to the field of mullite, which would be interesting for ceramic products, since mullite has a low coefficient of thermal expansion between 20 and 100°C of 5.2x10^ ^{6} °C^{-1}, as well as a low creep rate, good thermal and chemical stability and good mechanical strength. It can therefore be said that these residues are interesting for making ceramic products.

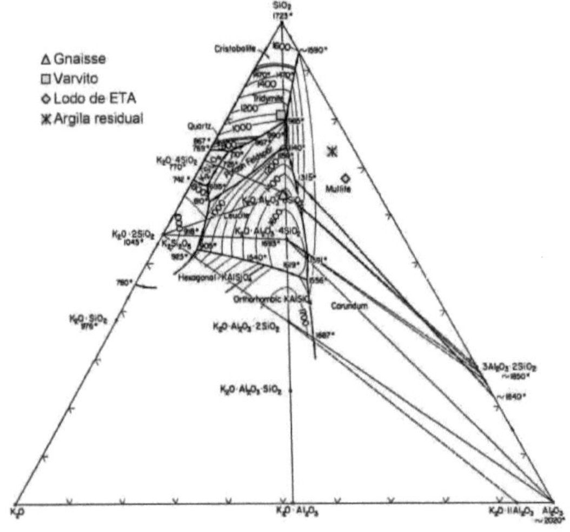

Figure 4.1: Phase diagram showing the positioning of in natura waste.

To determine the process parameters, the chosen phase diagram was analyzed. Within the area delimited by the *raw* waste *in* the diagram, three temperatures are found: 985, 1140 and 1315°C. Generally, in the ceramics industry for traditional products, temperatures above 1200°C are not used, because the higher the temperature, the greater the energy expenditure and consequently the higher the cost. Therefore, based on the temperatures found in the phase diagram, 1100 and 1150°C were chosen for sintering the products. With regard to the choice of stage, it should be borne in mind that longer times favor equilibrium, but the ceramics industry currently uses increasingly faster cycles. Therefore, based on these factors, we opted for a level of 40 minutes.

4.1.2. Mineralogical Analysis (XRD)

The X-ray diffractometry technique was used on the waste used in this work in order to identify the mineralogical phases present in the raw materials.

Figures 4.2 and 4.3 show the X-ray diffractograms of the four residues used as raw materials (gneiss, varvite, WTP sludge and clay). The analyses were carried out on the waste in natura (at room temperature), and sintered at 1100 and 1150°C. In these diffractograms, the intensity of the most intense peaks was cut off for better visualization, avoiding overlapping peaks.

Gneiss is a rock whose mineralogical composition is dominated by quartz, feldspars and mica. The diffractogram of this rock shows that it is rich in quartz (SiO_2), potassium feldspar ($KAlSi_3O_8$, microcline) and sodium feldspar ($NaAlSi_3O_8$, albite). In addition to these minerals, there are indications of the presence of hematite (Fe_2O_3) and magnetite (Fe_3O_4). The crystalline phases detected by XRD in the varvite sample were quartz and albite, as well as traces of dolomite ($CaMg(CO_3)_2$), clinochlore ($(Mg,Fe)_5Al(Si_3Al)O_{10}(OH)_8$) and forsterite ($Mg_2SiO_4$).

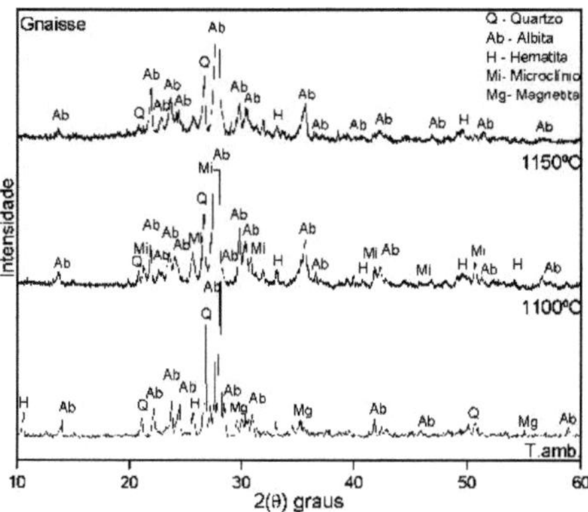

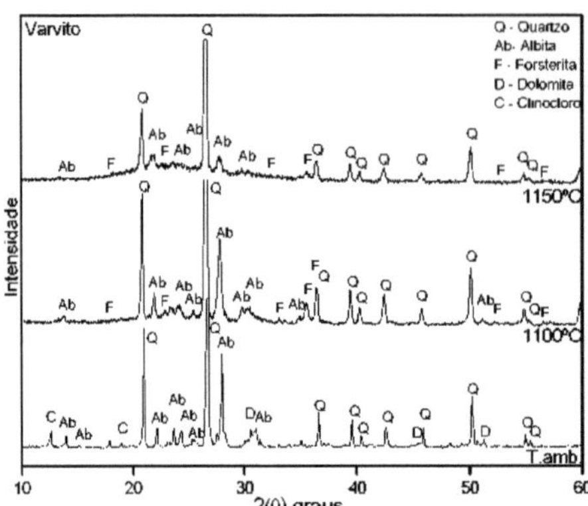

Figure 4.2: XRD of gneiss and varvite.

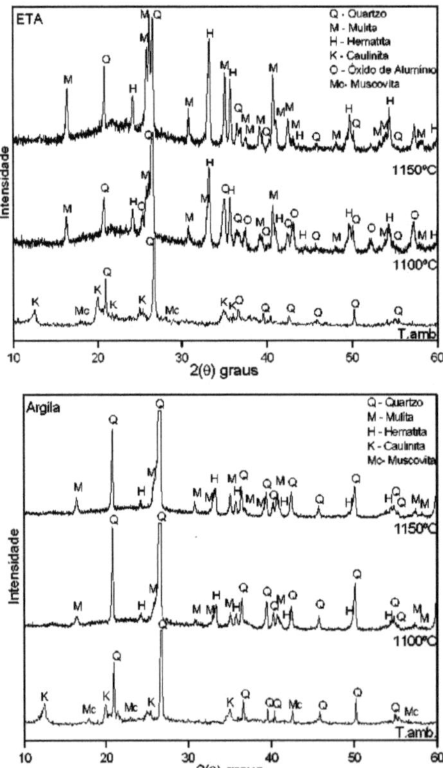

Figure 4.3: XRD of WTP sludge and clay.

The spectra of WTP sludge and clay at different temperatures were similar, since WTP sludge comes from rivers and contains a large amount of organic matter and is of clay origin. At room temperature, these two residues showed peaks of quartz, muscovite ($KAl_2(Si_3Al)O_{10}(OH,F)_2$) and kaolinite ($Al_4(SiO_{10})(OH)_8$), which is the main constituent of kaolin or clay. As the temperature rises, mullite peaks are formed by the transformation of kaolinite. The chemical composition of these two residues is very close. Therefore, at 1100 and 1150°C they both showed similar peaks, such as quartz, mullite ($3Al_2O_3 \cdot 2SiO_2$) and hematite. In addition, the WTP sludge spectra showed more intense hematite peaks.

4.1.3. Thermal Analysis

The thermal behavior of the residues was analyzed using the DTA/TG curves shown in Figure 4.4. These analyses were only carried out up to a temperature of 1000°C so that the samples would not stick to the crucible. The DTA analysis of the WTP sludge showed an endothermic peak at approximately 120°C due to moisture loss. The release of organic matter was also observed at around 240 to 500°C (exothermic), followed by an endothermic peak at 560°C referring to the dehydroxylation of kaolinite and an exothermic peak at around 620°C corresponding to the decomposition of sulphates (introduced at the water treatment plant in order to flocculate the particles). TG analysis reveals a total mass loss of this residue of approximately 20%.

The DTA graph for varvite shows a peak in moisture loss at 60°C. The thermal decomposition of clinochlore is seen during the outflow of water from the octahedral sheets which gives rise to an endothermic peak at ≈ 580°C and the outflow of water between the 2:1 layers gives rise to a second endothermic peak at around 850°C followed by an exothermic rise at around 870°C corresponding to the phase transformation giving rise to spinel, forsterite and enstatite [Villieras et al., 1994]. A total mass loss of 5.8% was detected in the TG curve of the varvite sample.

The DTA thermal analysis carried out on the gneissic rock crushing residue revealed two endothermic peaks, one at approximately 100°C, which is related to the elimination of moisture and, at 808°C, related to the decomposition of carbonates. This analysis also showed an exothermic peak at around 480°C, referring to the probable combustion of organic material. The mass loss detected by TG analysis was 3%.

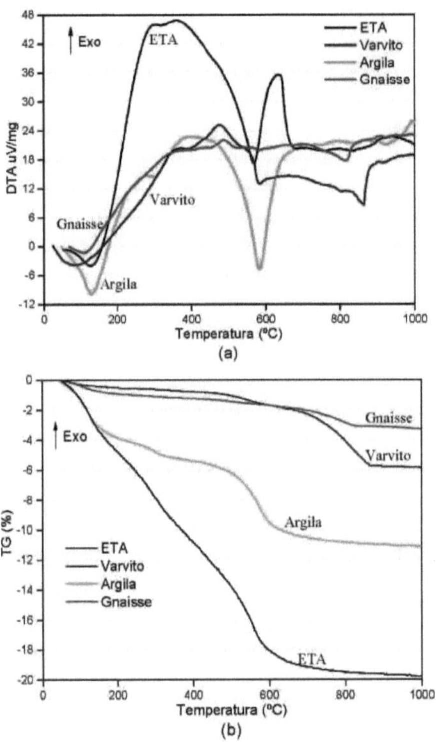

Figure 4.4: Thermal analysis of the residues under study (a) DTA and (b) TG.

4.1.4. Optical dilatometry

To understand the thermal behavior of the waste, the technique of optical dilatometry was used. The curves obtained are shown in Figure 4.5.

Analyzing the curve obtained, it can be seen that the waste that showed the greatest shrinkage (negative expansion) was WTP sludge, followed by clay. This observation can be better seen in Figure 4.5b, where the temperatures used, 1100 and 1150°C, are highlighted. The WTP sludge showed 6% shrinkage at 1100°C and 10.2% shrinkage at 1150°C, while the clay showed around 1.3% shrinkage at 1100°C and 2.2% shrinkage at 1150°C. The greater shrinkage of the WTP sludge is justified by its 12% loss on ignition, as shown in the XRF analysis. The varvite and gneiss residues show respectively 1.3% and 1.2% shrinkage at 1100°C and 1.0% and 1.1% shrinkage at 1150°C. However, these results are obtained without the presence of a sintering plateau, which would change the results. However, this analysis was important for identifying the maximum working temperature of the residues, which is 1150°C. The optical dilatometry test was used until the residues were completely melted.

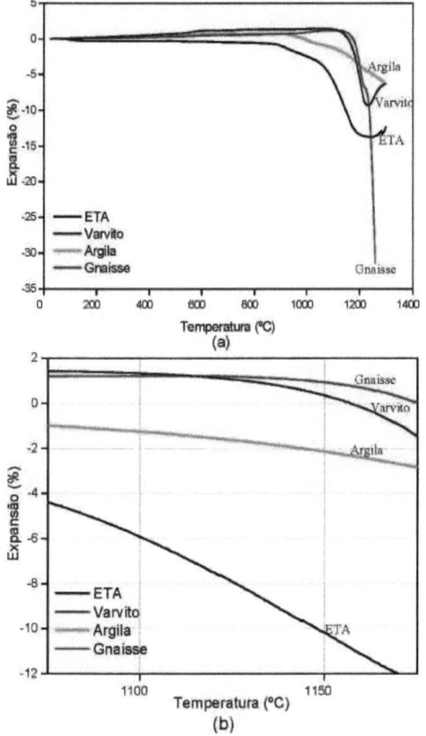

Figure 4.5: Thermal expansion curve of the waste obtained by optical dilatometry (a) the entire curve and (b) detail of the study area (1100 and 1150°C).

4.1.5. Particle Size Distribution

The determination of the particle size distribution of the samples was carried out on the samples as received, without any prior comminution process. The results shown in Table 4.2 show that 100% of the particles in all the residues are below 71 μm and 90% of all the residues are below 42 μm.

Table 4.2: Particle size distribution by laser diffraction.

Particle distribution ranges (μm)

Waste	10% below	50% below	90% below	100% below

	(µm)	of (µm)	of (µm)	of (µm)
Gneiss	1,37	13,37	42,10	71,00
Varvito	1,04	7,87	21,53	36,00
WTP sludge	0,87	4,37	17,67	36,00
Residual clay	0,86	5,30	21,03	45,00

This analysis showed that the residues are in a fine (50% is between 4 and 13 µm) and narrow particle size range, which favours good homogenization and better packing during forming. This fine particle size distribution brings the advantage of not needing comminution processes, thus making the process cheaper.

4.2. CHARACTERIZATION OF FORMULATIONS

Using the points marked on the phase diagram, which correspond to the *raw* waste, four formulations were selected (Figure 4.6).

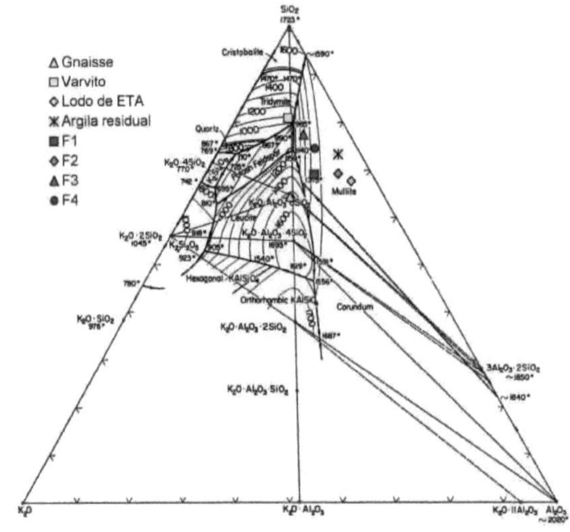

Figure 4.6: Phase diagram showing the positioning of the established formulations and the raw waste.

The percentages of the four residues present in a given formulation were chosen based on the position of the formulation within the phase diagram, in order to cover the largest possible area within the delimited space. The percentage of each residue present in the formulations is shown in Table 4.3.

Table 4.3: Established formulas.

Waste	Formulations (%)			
	F1	F2	F3	F4
Gneiss	45	10	5	10
Varvito	5	5	65	40
ETA	10	65	10	20
Residual clay	40	20	20	30

Once the formulations had been chosen, test specimens were evaluated for water absorption and density in preliminary pressing tests. These tests were carried out to determine whether the process parameters chosen using the phase diagram were really the most suitable. In the final stage, in which the samples were formed by extrusion, the samples were analyzed using techniques such as XRD, optical dilatometry, plasticity, linear shrinkage, water absorption and mechanical flexural strength (3 points).

4.2.1. Preliminary Tests

In order to check whether the working temperatures chosen from the phase diagram (1100 and 1150°C) are the most suitable for the job, some preliminary tests were carried out. Pressed inserts were used in these tests. Different forming processes give different properties to the materials, but these tests were carried out in order to obtain a prediction of the material's behavior. The tests carried out were water absorption, geometric, *bulk*, pycnometric and apparent densities (calculated by the Archimedes method in water) for samples sintered at 900, 950, 1000, 1050, 1100 and 1150°C. The specimens were made by uniaxial pressing at a pressure of 50 MPa and sintered for 40 min at a heating rate of 5°C/min.

4.2.1.1. Water absorption

For the water absorption test, 5 pressed samples were used for each temperature and formulation. The results obtained are shown in Table 4.4. As the sintering temperature increases, the water absorption value obtained decreases. The material becomes more compact, with less open porosity.

Table 4.4: Water absorption of formulations at different temperatures obtained by pressing

Temp.	F1	F2	F3	F4

(°C)	σ (%)	D.P.	σ (%)	D.P.	σ (%)	D.P.	σ (%)	D.P.
900	16,22	0,49	21,44	0,85	21,48	0,96	20,26	0,36
950	15,24	0,25	20,08	0,72	21,48	0,34	19,34	0,78
1000	14,17	0,46	14,28	0,85	20,68	0,81	18,46	0,98
1050	13,50	0,93	8,62	0,50	20,85	0,69	17,85	0,62
1100	9,58	1,10	4,49	0,43	6,62	1,30	8,19	1,07
1150	0,91	0,21	0,75	0,31	0,99	0,45	0,82	0,46

σ - Mean; SD - Standard Deviation.

It can be seen that from the sintering temperature of 1100°C, all the formulations show water absorption within the specifications normally used in ceramic tiles (up to 10%). At 1150°C, all the samples showed water absorption of less than 3% and could be classified as stoneware. Figure 4.7 shows the decrease in water absorption with increasing temperature.

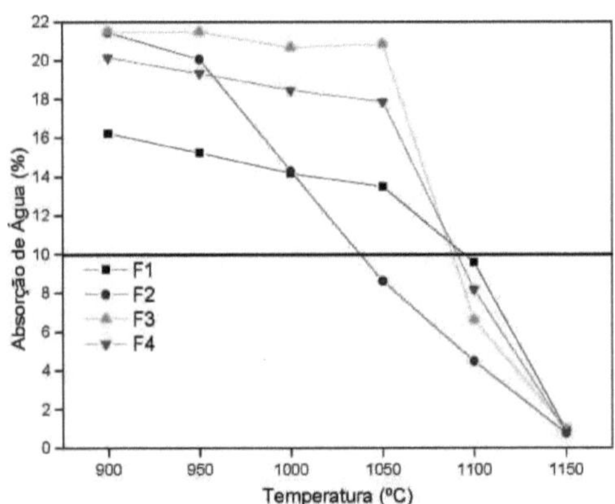

Figure 4.7: Water absorption of pressed and sintered samples from 900 to 1150°C.

Although forming by pressing gives different properties to the materials, this test shows that the most suitable temperatures for the study are 1100 and 1150°C. At these temperatures, the low value of water absorption in the samples means that the material is more consolidated, which consequently results in greater mechanical strength.

4.2.1.2. Density

The pycnometric, Archimedean, *bulk* and geometric densities were measured, Figure 4.8 and 4.9.

Pycnometric density is obtained using the powder of ground samples, not taking pores into account. Archimedes density only includes open porosity, but not closed porosity. The *bulk* density calculation takes into account all types of porosity present in the sample. Geometric density is measured from the dimensions and mass of the specimens.

In all the formulations, the densities of the samples behaved similarly. The pycnometric and Archimedes densities decreased as the sintering temperature increased, while the geometric and bulk densities increased as the sintering temperature increased.

The pycnometric density decreased with increasing temperature, as it only corresponds to the density of the solid. As the temperature rises, a phase change can occur and the sample can crystallize, becoming denser.

The Archimedean density decreases due to the change in crystalline phase, combined with the increase in closed porosity, resulting in a lower density than that obtained by pycnometry. According to the graphs, up to 1100°C the density of the solid is decreasing, and above 1100°C, the increase in closed porosity has a greater influence on the drop in density.

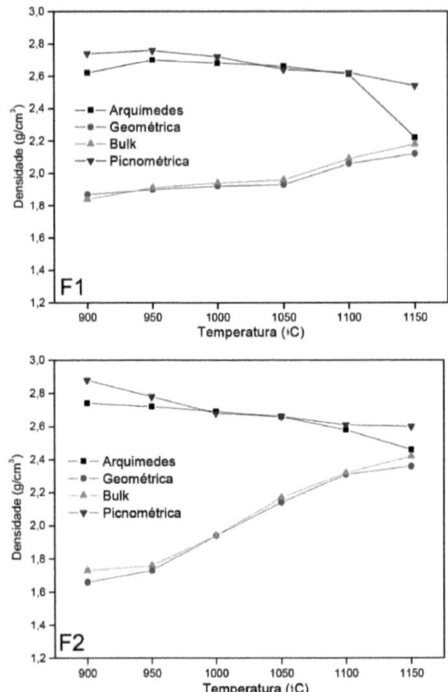

Figure 4.8: Density of samples F1 and F2 sintered from 900 to 1150°C.

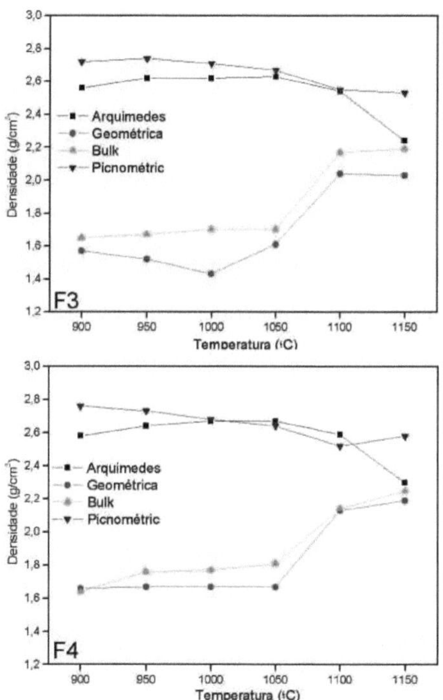

Figure 4.9: Density of samples F3 and F4 sintered from 900 to 1150°C.

The results of the *bulk* and geometric densities can be analyzed together. The densities take into account the open and closed pores of the samples. These densities gradually increased up to 1150°C, as the amount of open pores decreased, which can be confirmed by the water absorption of the samples. However, above 1100°C, the material forms a liquid phase, causing the pores to be trapped inside the sample, resulting in an increase in closed porosity, as can be seen in the scanning electron microscopy images, Figures 4.19 to 4.22.

Through these analyses, it was found that the density of the samples is between 2.03 and 2.61 g/cm^3 for temperatures of 1100 and 1150°C, values normally found in ceramic tiles.

4.2.2. Plasticity

The liquidity limit test indirectly measures the shear strength of the sample for a given moisture content, through the number of blows required for the sample to slide. Using the Casagrande method, it was possible to determine the moisture content, thus obtaining a pair of values, "moisture content x number of blows", which will define a point on the creep graph. Repeating this procedure for different moisture contents (40, 30, 20 and 10± 3 blows) enabled the graphs shown in Figures 4.10 and 4.11 to be constructed.

It has been agreed that, in the Casagrande test, the moisture content corresponding to 25 blows needed to close the groove is the liquidity limit. The liquidity limit of a sample is the moisture content that separates the liquid state of consistency from the plastic and for which the sample has a low shear strength.

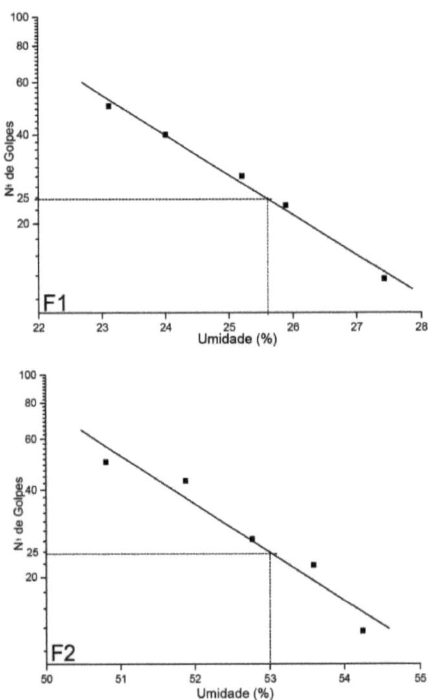

Figure 4.10: Liquidity limit of the formulations using the Casagrande method for F1 and F2.

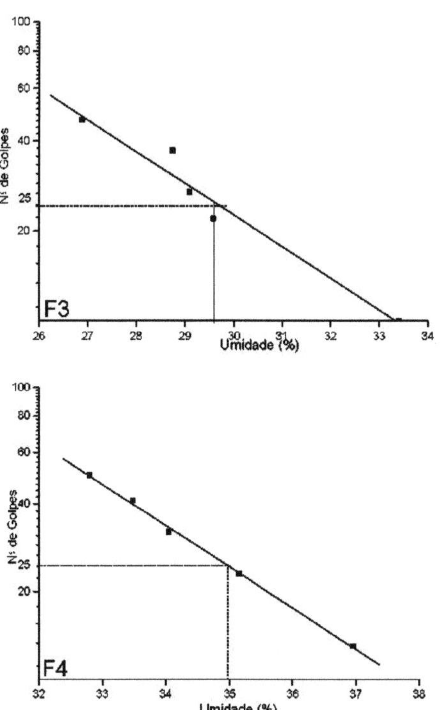

Figure 4.11: Liquidity limit of the formulations using the Casagrande method for F3 and F4.

The results obtained for the plasticity characteristics using the Casagrande apparatus to find the liquid limit and the roller method to find the plastic limit are shown in Table 4.5. The LL values obtained range from 25.7% to 53.2% of added water, while the LP values range from 9.1% to 30.3%. The IP is found by subtracting the LP from the LL.

Table 4.5: Plasticity characteristics of the established formulations.

Samples -	Plasticity indices (%)		
	LL	LP	IP
F1	25,7	9,1	16,6
F2	53,2	30,3	22,9
F3	29,6	15,2	14,4
F4	35,0	15,6	19,4

The plasticity index (PI) determines the plasticity of a sample, so the higher the PI, the more plastic the material. According to the classification of Caputo [1998] cited in section 2.2.1.3, with the exception of sample F3, which had an index of 14.4% and was classified as moderately plastic, all the formulations can be classified as highly plastic, as they had an IP > 15.

4.2.3. Mineralogical Analysis (XRD)

The X-ray diffractometry technique was used in this work to identify the mineralogical phases of the formulations and to characterize the ceramic materials developed in terms of the quantitative phases present.

Figures 4.12 and 4.13 show the X-ray diffractograms of the formulations (F1, F2, F3 and F4) sintered at 1100°C and 1150°C, with a 40 min plateau. The variations that occurred as the sintering temperature increased can be seen by superimposing the diffractograms.

Because they were produced from waste and not pure components, the formulations developed showed several different X-ray patterns.

The F1 formulation is composed predominantly of gneiss (45%) and clay (40%). Gneiss is a rock with a predominant mineralogical composition of quartz, feldspars and mica. The diffractogram, Figure 4.12, shows peaks of the most common feldspars such as albite ($NaAlSi_2O_8$), anorthite ($CaAl_2Si_3O_8$) and orthoclase ($KAlSi_3O_8$). Microcline ($KAlSi_3O_8$) is also a potassium feldspar which differs from orthoclase by its triclinic and pinacoidal crystallography, while orthoclase has monoclinic and prismatic crystallography.

As all the residues have quartz as a predominant component in their composition, quartz peaks were expected. The mullite peaks formed as a result of the calcination of the clay and the WTP sludge present in the formulation by transforming the kaolinite.

When the F1 formulation is sintered at 1150°C, peaks of quartz, albite, mullite and hematite are evident. It can be seen that at 1150°C quartz decreased the intensity of its main peaks. Peaks corresponding to hematite also appeared in the diffractogram of the F1 formulation at 1150 °C. Hematite is a mineral widely distributed in rocks of all ages and forms the most abundant iron ore, occurring in metamorphic rock deposits such as gneiss. The F1 formulation contains around 5% Fe2O3 in its composition, which comes from the residues that make it up. Hematite, Fe_2O_3, is a component considered an impurity that gives the paste its red, brown and yellow colors.

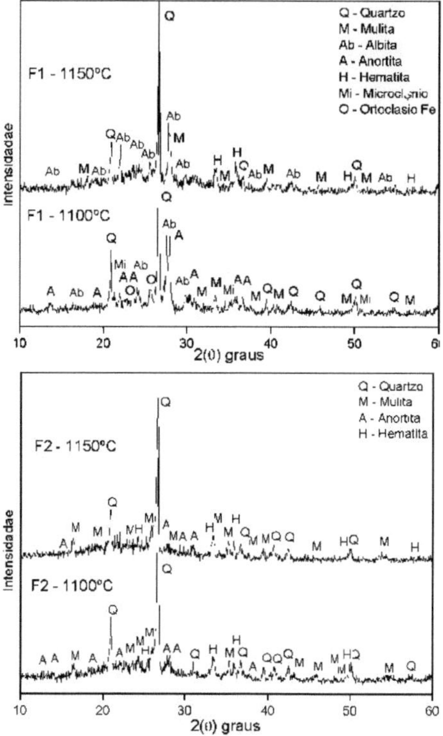

Figure 4.12: Diffractograms of formulations F1 and F2 at 1100°C and 1150°C.

Formulation F3 is composed predominantly of varvite (65%) and clay (20%). In addition to the quartz peaks, the diffractogram shows peaks of the feldspars albite and anorthite. Peaks of the magnesium silicate known as enstatite ($Mg_2(Si_2O_6)$) were also identified. This silicate may have formed because among the residues, varvite has the highest amount of magnesium at 1.74% and there is plenty of quartz in the composition for this combination.

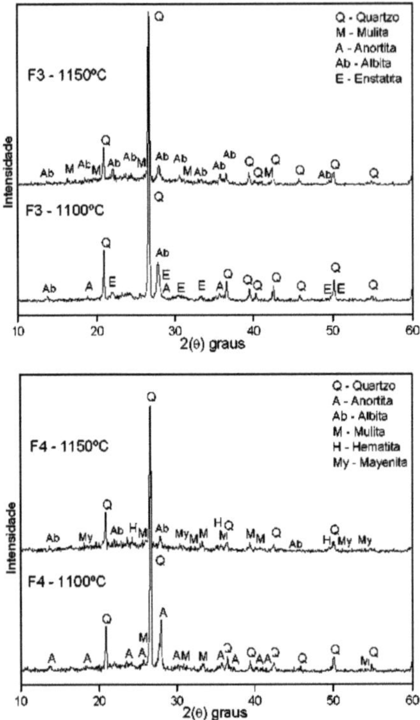

Figure 4.13: Diffractograms of formulations F3 and F4 at 1100°C and 1150°C.

When formulation F3 is sintered at 1150°C, only quartz, albite and mullite peaks are observed. Once again, the intensity of the quartz peaks decreased with increasing temperature, as did the peaks corresponding to albite.

In formulation F4, the main residues are varvito (40%) and clay (30%). This has the same major constituents as formulation F3, but with a lower percentage of varvite residue and a higher percentage of clay. The diffractogram of formulation F4 at 1100°C showed peaks of quartz, anorthite and mullite.

When the F4 formulation is sintered at 1150°C, peaks of quartz, mullite, albite, a calcium aluminate ($Ca_{12}Al_{14}O_{33}$) known as mayenite and hematite are observed. Again, the intensity of the quartz peaks decreased with increasing temperature.

The diffractograms of the four formulations showed peaks of mullite, as predicted by the phase diagram, but due to the amount of components present in these four residues individually, there were also peaks of other phases.

4.2.4. Specific Surface Area Analysis

The value obtained for the raw material samples, using the BET technique, was 18.6 m^2/g for the F1 formulation. Sample F2 had the highest specific surface area among the samples, 24.9 $m^{(2)}$)/g. Samples F3 and F4 had 11.4 and 16.4 m^2/g respectively. The formulations had a relatively high specific surface area due to the low particle size distribution of the residues that make up the formulations.

4.2.5. Thermal Analysis

The thermal behavior of the formulations was analyzed using the DTA/TG curves shown in Figure 4.14. The DTA curve of the F1 formulation shows the release of free water at 110°C, the loss of organic matter from approximately 280 to 500°C, an endothermic peak at 570°C corresponding to the release of water from the clays and a slight exothermic peak at around 982°C (with no loss of mass) which corresponds to the start of mullite formation (as identified by XRD). The corresponding mass losses are visible in the TG curve, with a total mass loss of 8%.

In the DTA analysis carried out on formulation F2, the same peaks were identified as those found in formulation F1 (loss of water at 110°C, of organic matter at approximately 250 to 520°C and of water constituting the clays at 570°C). However, as this formulation contains 65% WTP sludge, there was an exothermic peak at around 610°C corresponding to the decomposition of sulphates (introduced at the water treatment plant with the aim of flocculating the particles). Although mullite was identified in the corresponding XRD, there was no mullite formation peak. The total mass loss detected by the TG curve of this formulation was 16%.

The same first three peaks of the F3 formulation are also identified at detected in the F1 and F2 formulations. In addition to these three peaks, an endothermic peak was also identified at approximately 820°C, probably corresponding to the decomposition of carbonates (dolomite) present in the varvite in this 65% formulation.

It is also possible to observe a slight exothermic peak in the curve of formulation F3 at around 980°C (without mass loss) which corresponds to the beginning of mullite formation (as identified by XRD). This formulation showed a total mass loss detected by the TG curve of 8%.

The analysis carried out on formulation F4 is very similar to that carried out on formulation F3. However, the peak corresponding to the decomposition of carbonates at approximately 800°C is less pronounced, as the amount of varvite was reduced from 65% to 40%, thus reducing the amount of carbonates present in the formulation. Although there is mullite in its corresponding XRD, a mullite formation peak is not visible either. The total mass loss detected by the TG curve for this formulation was 10%.

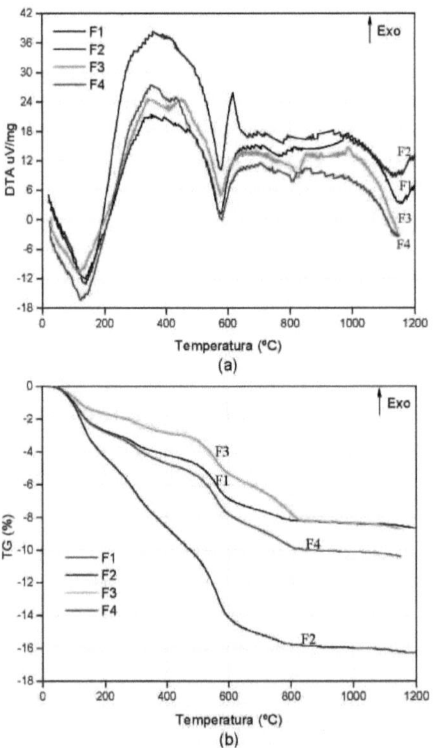

Figure 4.14: Thermal analysis of the formulations (a) DTA and (b) TG.

4.2.6. Optical dilatometry

For a more detailed study of the thermal expansion of the formulations, the optical dilatometry technique was used. The curves obtained are shown in Figure 4.15.

It can be seen from the expansion curve that the greatest shrinkage is found in F2. This formulation showed 7.5% shrinkage at 1150°C and 4% shrinkage at 1100°C. Formulation F2 is the sample with the highest amount of WTP sludge, which in turn shows around 12% loss on ignition as shown in the chemical analysis. The other formulations show 1 to 2% shrinkage at 1150°C and 0 to 0.6% shrinkage at 1100°C.

However, this analysis did not apply thresholds, and the thermal behavior of materials is different when a thermal cycle has thresholds. This analysis is important, however, for estimating the maximum working temperature of the materials.

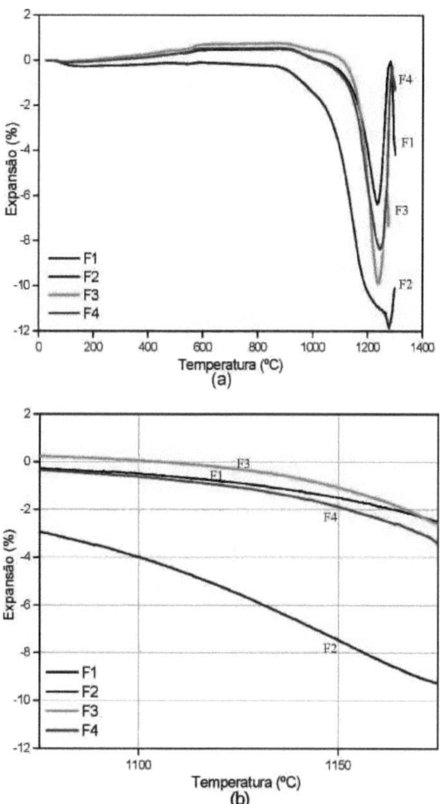

Figure 4.15: Thermal expansion curve of the formulations obtained by optical dilatometry (a) the entire curve and (b) detail of the study area (1100 and 1150°C).

4.2.7. Linear Firing Retraction

The F1 formulation samples showed shrinkage of 4.61 and 6.16% at 1100°C and 1150°C respectively. The F2 samples showed greater shrinkage at both 1100°C and 1150°C due to the higher concentration of organic matter from the WTP sludge. The results obtained from the linear shrinkage tests are shown in Table 4.6.

Table 4.6: Linear shrinkage of the different extruded and sintered formulations.

Temp (°C)	F1		F2		F3		F4	
	σ (%)	D.P.	σ (%)	D.P.	σ (%)	D.P.	σ (%)	D.P.
1100	4,61	0,50	14,01	0,38	10,41	0,28	10,84	0,54
1150	6,16	0,21	13,19	0,79	2,51	0,81	6,00	1,19

σ - Mean; P.D. - Standard Deviation.

Formulation F3 showed 10.41% at 1100°C and 2.51% at 1150°C. Generally, the higher the temperature, the greater the linear shrinkage at firing, but formulation F3 showed the opposite results. This can be explained by the deformation that occurred in the samples sintered at 1150°C. These samples showed swelling, distortion and a large amount of porosity throughout, giving the appearance of boiled material, thus masking the results of the analysis. This effect can also be seen in the water absorption and mechanical strength of these samples.

Figure 4.16, which shows the graph of the shrinkages with their respective standard deviations, shows that the F4 samples sintered at 1150°C also suffered the same effect as the F3 samples at 1150°C, but to a lesser extent. These samples did not show pores to the same extent as the F3 samples, but the shrinkage of this formulation at 1150°C was also masked by the swelling of the specimens.

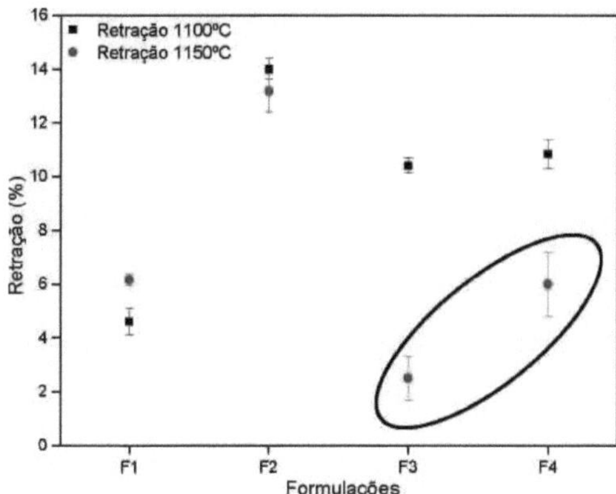

Figure 4.16: Linear firing retraction of extruded samples.

The formation of a large number of pores during the sintering of formulations F3 and F4 at 1150°C may be due to the release of gases during the reduction of hematite (Fe_2O_3) into iron monoxide.

4.2.8. Water absorption

The water absorption results of the extruded samples are shown in Table 4.7. The water absorption test was applied to ten samples sintered at 1100 and 1150°C.

The F1 formulation showed 9.28% water absorption at 1100°C and the samples sintered at 1150°C showed 0.56% absorption. The F2 samples showed very similar water absorption at both 1100 and 1150°C.

Table 4.7: Water absorption of extruded and sintered formulations at 1100 and 1150°C.

Temp (°C)	F1		F2		F3		F4	
	σ (%)	DP	σ (%)	DP	σ (%) D	DP	σ (%)	DP
1100	9,28	0,96	0,74	0,22	0,46	0,13	1,95	0,63
1150	0,56	0,17	0,84	0,13	1,27	0,46	5,24	1,43

σ - Mean; P.D. - Standard Deviation.

The water absorption result obtained by the F3 formulation was 0.46% at 1100°C. When F3 was sintered at 1150°C, the samples swelled and showed porosity throughout, thus masking the water absorption result for these samples. The same occurred with the F4 samples sintered at 1150°C, although the distortion and porosity developed with temperature were lower in these samples.

The water absorption of all the samples is below 10%, which is in line with the range used in ceramic tiles. With the exception of F1, all the formulations sintered at 1100°C are within the working range specified in the standard for the production of stoneware and semi-stoneware (stoneware: 0.5<AA≤ 3 and semi-stoneware: 3<AA≤ 6). The results obtained can be better visualized in Figure 4.17, which shows the graph of water absorption with their respective standard deviations.

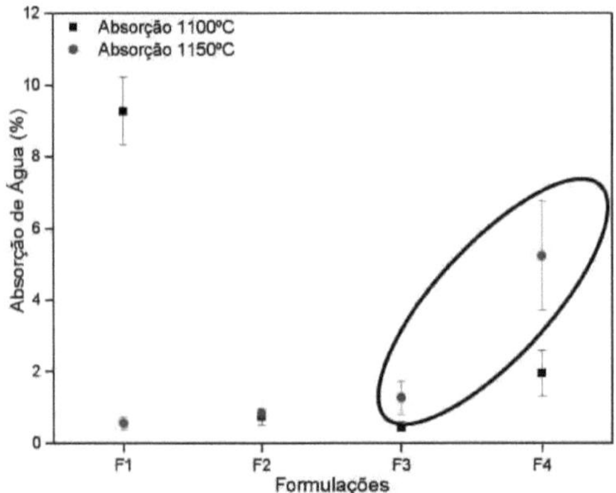

Figure 4.17: Water absorption of extruded and sintered samples at 1100 and 1150°C.

It can be seen from the graph that the greatest difference between the water absorptions at the different temperatures is found in the F1 formulation, and that the absorption results found for the F2 samples are so close that they overlap in the figure. F3 also has very close absorption values for the samples sintered at 1100 and 1150°C and the F4 samples sintered at 1150°C have the highest standard deviation.

4.2.9. Mechanical Flexural Strength

For ceramic tiles, the most widely used test for measuring mechanical strength is the flexural test. Flexural strength is defined as the maximum tensile stress at break.

The sintering temperature used has a significant influence on the mechanical strength of the formulations, as can be seen in Table 4.8.

Table 4.8: Mechanical flexural strength of extruded formulations

		F1	**F2**	**F2**		**F4**			
Temp (°C)		°	**D.p. Λ< σ**		**D.p. Λ< σ**		**D.p. Λ< σ**		**D.P.**
		MPa	**MPa**	**MPa**		**MPa**			
1100	27,08	3,71	46,90	4,48	58,32	5,11	40,88	5,21	
1150	37,23	4,82	55,02	2,54	21,32	2,66	26,26	3,24	

σ - Mean; P.D. - Standard Deviation.

The results showed that the heat treatment carried out at 1150°C caused the appearance of cracks and internal porosity, due to the formation of the liquid phase, causing the mechanical strength to decrease in the samples made with the F3 and F4 formulations. The action of the firing temperature on the mechanical flexural strength of the samples can be better seen in Figure 4.18.

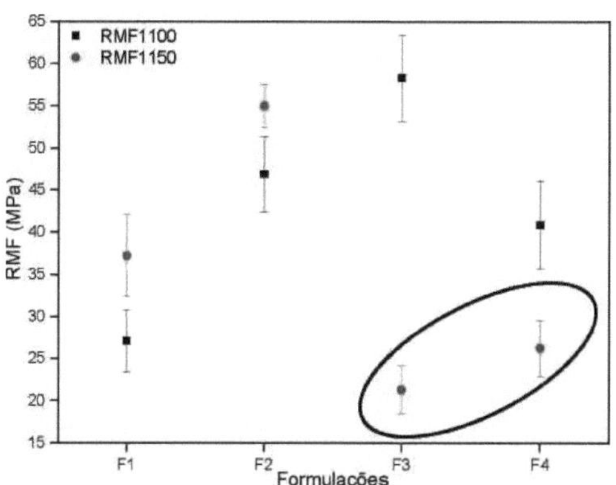

Figure 4.18: MFR of the extruded and sintered samples at 1100 and 1150°C.

The samples made with the F1 and F2 formulations also changed with the high temperature, but to a lesser degree. The F1 samples showed some internal porosity and the F2 samples remained compact, but internal cracks appeared. F1 and F2 correspond to higher mechanical flexural strengths at 1150°C than at 1100°C.

The MFR of all the samples tested was in the range of 21.32 to 58.32 MPa, which is within the requirements of the ISO 13006 standard for semi-stoneware (22 to 35 MPa) and, with the exception of the F1 formulation, even for stoneware (35 to 45 MPa).

4.2.10. Analysis of Fracture Surface Microstructures

The morphological characterization of the formulations was carried out using scanning electron microscopy on the fracture surfaces of the samples subjected to mechanical bending tests. Figure 4.19 shows the micrographs of two different regions at 50 and 500 times magnification of the F1 formulation, sintered at 1100 and 1150°C; and the qualitative chemical analysis of a selected region obtained by energy dispersive microanalysis (EDS).

An important feature of microanalysis is the possibility of obtaining a compositional map of the region under observation, allowing electron microscopy to be correlated with detailed compositional information. In the compositional map, the red color corresponds to the presence of silica and the green and blue colors correspond respectively to the presence of alumina and potassium. The chemical composition of a region of the formulation shows that the sample is homogeneous, with no concentration of elements in any particular part.

It can be seen that the F1 sample sintered at 1100°C is more compact, showing less porosity and irregularity than the sample sintered at 1150°C.

The F2 formulation showed great linear shrinkage in its samples, resulting in the formation of cracks and micro-cracks that are visible in the micrographs shown in Figure 4.20.

The F3 formulation is well consolidated at 1100°C, but at 1150°C this formulation has developed a large number of porosities and micro-porosities visible even to the naked eye, Figure 4.21.

The F4 formulation at 1100°C is well consolidated and shows a small amount of micro-pores while at 1150°C this formulation has developed a large amount of porosity affecting its mechanical properties and its visual characteristic, Figure 4.22.

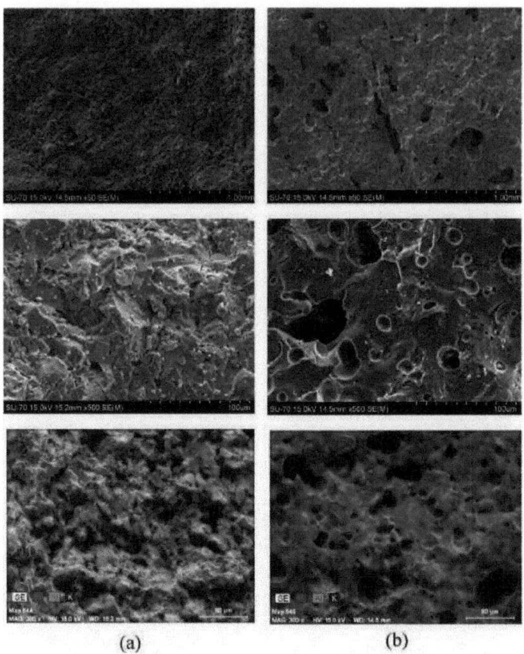

Figure 4.19: Fracture surfaces of the F1 formulation (a) 1100°C and (b) 1150°C.

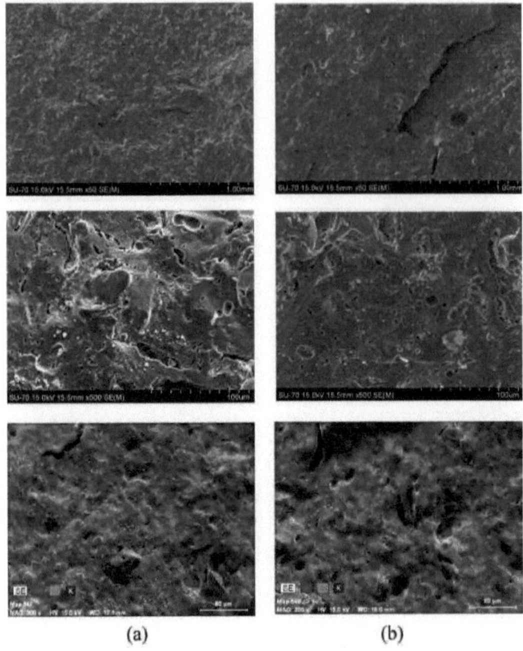

Figure 4.20: Fracture surfaces of formulation F2 (a) 1100°C and (b) 1150°C.

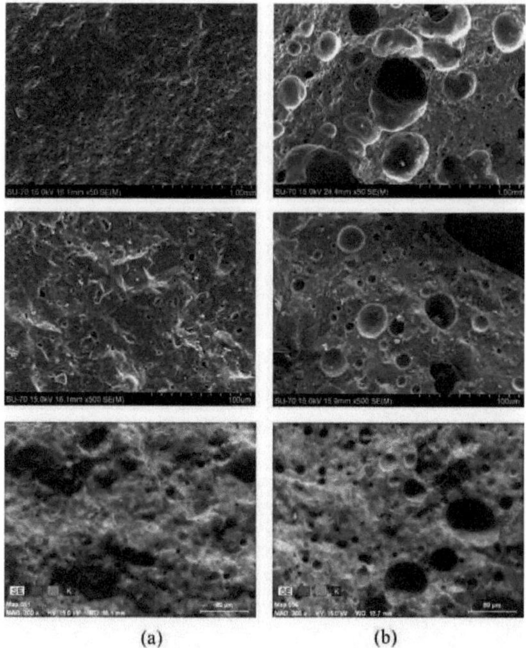

Figure 4.21: Fracture surfaces of formulation F3 (a) 1100°C and (b) 1150°C.

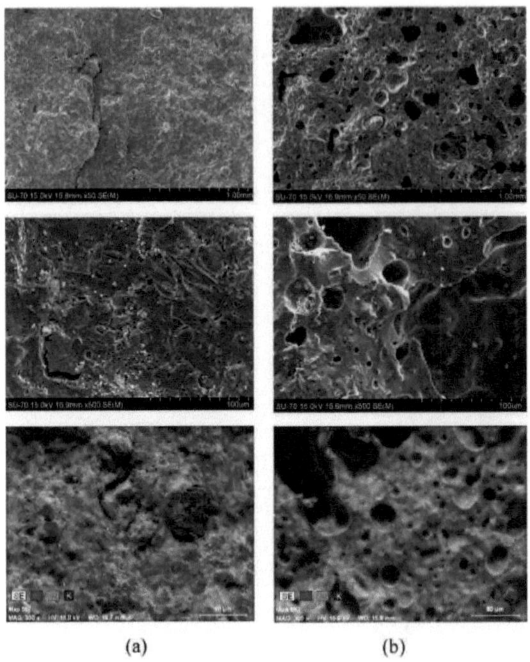

4.2.12. 4.22: Fracture surfaces of formulation F4 (a) 1100°C and (b) 1150°C.

The fracture surface of the samples can also be seen in Figure 4.23, where you can see the porosity developed in samples F1, F3 and F4 and the cracks caused in samples F2 at 1150°C. It can also be seen that F1 and F2 had some aggregate points in their samples, due to poor homogenization of the raw materials as a result of using a piston extruder.

Amostras	F1	F2	F3	F4
1100°C				
1150°C				

4.2.13. 4.23: Fracture surface of extruded and sintered samples at 1100 and 1150°C.

However, the high temperature affected the appearance of the samples, distorting them and leaving them irregular and deformed. In the ceramics market, the aesthetic appearance of a coating is the first thing to be analyzed in the event of a purchase. For this reason, even though they met the market specifications, some of the formulations were discarded for processing, as shown in Figure 4.24.

When comparing the results obtained by the F1 formulation at 1100°C and 1150°C, it can be seen that at 1150°C, there was a significant decrease in water absorption (from 9.28% to 0.56%), an increase in shrinkage of around 2% and an increase in strength from 27.08 MPa to 37.23 MPa. These data show that the F1 formula, when sintered at 1100°C, did not achieve complete consolidation, requiring a higher temperature, which can be seen with the naked eye. For this reason, the F1 formulation sintered at 1100°C was discarded as unsuitable for the formulation of ceramic coatings.

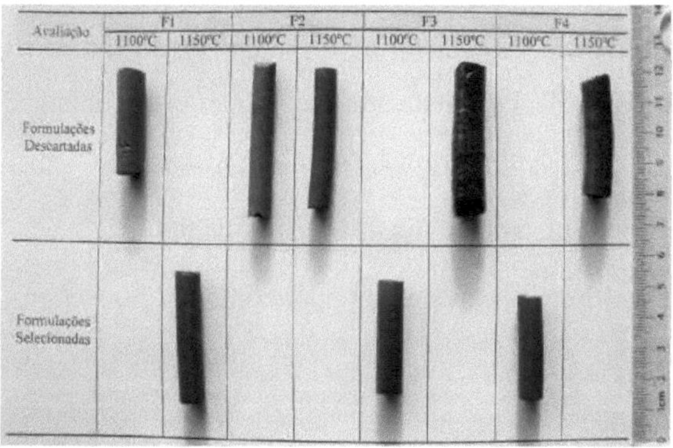

Figure 4.24: Evaluation of the most suitable formulations.

The results obtained for water absorption and mechanical flexural strength at both temperatures studied for the F2 formulation were very positive. However, the high percentage of linear shrinkage in these samples (14.01% at 1100°C and 13.19% at 1150°C), in addition to causing cracks to form, makes it difficult to control their dimensions, which would be a problem in mass production, so the F2 formulation was discarded for making ceramic tiles.

The F3 formulation at 1100°C showed adequate results in terms of water absorption (0.46%), linear shrinkage (10.41%) and mechanical flexural strength (58.32 MPa), as well as a compatible visual appearance, making it suitable for the production of ceramic tiles. However, at 1150°C this same formulation showed severe deformation of its specimens, as seen in Figure 4.24, and this temperature was discarded for processing purposes.

The results presented by the F4 formulation at 1100°C were also within the current standards, obtaining 1.95% water absorption, 10.84% linear shrinkage and 40.88 MPa mechanical flexural strength, as well as a good visual appearance, making this formulation suitable for the production of ceramic tiles. However, at 1150°C this formulation showed an increase in porosity and deformation of the samples, which led to a decrease in mechanical properties, and consequently this temperature was discarded.

4.3. PRELIMINARY ECONOMIC VIABILITY

The first thing to check when considering manufacturing a product on an industrial scale is the amount of raw material available, i.e. whether the amount of waste obtained is enough to meet the needs of a ceramics company on a monthly basis.

To estimate the amount of material needed, assume the following situation:

- Pieces measuring 240 x 115 x 10mm and weighing approximately 600 grams, such as those shown in Figure 4.25.
- Production capacity of 30 extruded parts per minute;
- The company operates for eight hours a day.
- Monthly production of +/- ~500,000 pieces.

As each formulation has a different percentage of the four residues used in its composition, the largest amount of the residue used was taken as the basis, regardless of the composition. The highest percentage of varvite used is in formulation F3 (65%), and according to the condition above, 170 tons/month of this material will be needed. The same is true of the WTP sludge, whose F2 formulation accounts for 65% of the composition. According to the condition imposed, 170 tons/month of this waste would also be needed. For gneiss, the highest percentage used of this material is 45% in the

F1 formulation, which means that 117 tons/month of this waste are required.

The amount of residual clay available from Eliane's stock alone is enough for 28 months of work. Yellowish and reddish clays can occur both above the light clay deposit and below and inside the light clay layer, in the form of intercalations or contaminated pockets. Thus, this clay can be found in many other deposits in the region.

WTP sludge is another waste that has sufficient quantity, as there are several other water treatment plants in the municipalities around Blumenau, which has four plants.

Figure 4.25: Example of a ceramic plate to be manufactured.

In the case of varvito, the amount of waste from just one company would not be enough for industrial-scale production. However, the Trombudo Central region has several other companies in the same sector that could supply the monthly consumption requirement.

For the gneiss residue, it would be necessary to stock up on the material before starting production, as this residue comes from just one company in the region. However, if it is necessary to look for another source of this waste, there are other gneiss processing companies in the state of Minas Gerais.

Generally speaking, if there is planning and time to stock up before production starts, all the waste used in the formulations has sufficient quantities for industrial-scale production, solely using waste as a source of raw materials.

The municipalities that supply the waste are located in a ceramics hub, which could make it feasible to set up a pilot plant for this project. The municipality chosen for this implementation would be Rio do Sul, due to its location, shown in Figure 4.26, showing its proximity to the waste supplying municipalities.

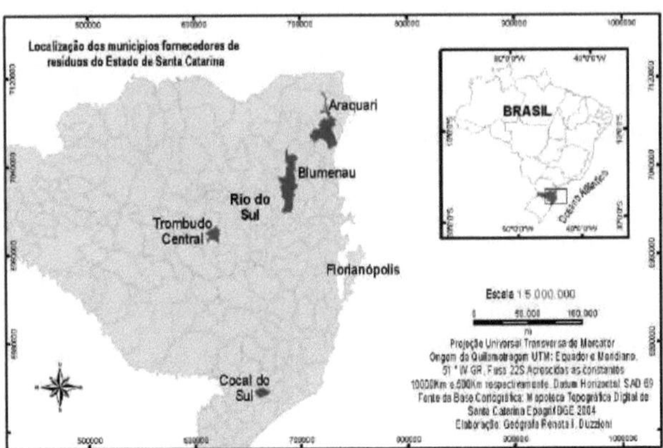

Figure 4.26: Municipalities where the companies supplying the waste are located and the location of a possible pilot plant.

The use of waste from settling ponds can bring an advantage to the extrusion process. If the waste is mixed as received, especially the WTP sludge, there would be no need to add water to the process, thus making it more economical. The materials would be homogenized, if necessary filter-pressed and then extruded.

Another saving in the process would come from not needing glazing, as these products would be classified as natural products, where the color of the waste would determine the color of the finished product. These coatings could be used on floors and walls in internal and external areas of commercial and residential premises .

5. CONCLUSIONS AND SUGGESTIONS

5.1 CONCLUSIONS

The results obtained confirm the possibility of valorizing and recycling mineral waste from the state of Santa Catarina as an alternative source of raw materials for the ceramic tile industry. These residues can be used as products with a high added value, with positive effects in environmental and economic terms. The product generated can be considered "environmentally friendly", and this information can even be used as a marketing strategy to sell these products.

After carrying out this work and analyzing the results obtained during each stage of the experimental procedure, the following final considerations can be made.

- The four residues studied constitute an attractive and renewable alternative source of ceramic raw materials.

- These residues used as alternative raw materials have the advantage of not needing comminution processes, since the way they are received they have 90% of their particles below 42 μm, with an average size of 13 μm, generating savings for the process.

- Another saving in the process would come from not having to glaze, as these products would be classified as natural products, with the color of the residues determining the color of the finished product.

- The phase diagram was successfully used to guide the choice of formulations while also helping to choose process parameters. In all cases, the diagram provided a basis for making predictions about the material's behavior under various service and processing conditions.

- The experiments on shrinkage, water absorption and mechanical resistance to bending indicated that the best firing cycle to use for these alternative raw materials is 1100°C with a 40 min landing time.

- Extrusion molding proved to be viable for processing these materials, as they showed sufficient plasticity, with plasticity indices above 15, which is considered highly plastic, not requiring additives, once again making the process cheaper.

- The results obtained from the water absorption analysis showed that formulations F2, F3 and F4 had an absorption rate of less than 2% at 1100°C, meeting the standard for the production of stoneware and semi-stoneware.

- The mechanical flexural strength of the formulations is in the range of 21 to 58 MPa, which is within the requirements of the ISO 13006 standard for semi-stoneware and, with the exception of the F1 formulation, even for stoneware.

- These coatings can be applied to floors and walls in interior and exterior commercial and residential premises.

5.2 SUGGESTIONS

Based on the results obtained in this research, some paths can be identified for its continuation:

- Developing a quality and particle size control system for the waste generated in the different processes directly at the company's plant.

- Optimize the mixing and extruding process of the formulations.

- Use the characterized waste for higher value-added uses.

- Test other times and temperatures in the sintering process, following the evolution through XRD.

- The use of a continuous (screw) extruder can improve the homogeneity of the mass and thus reduce defects and aggregates in the products obtained.

REFERENCES

BIBLIOGRAPHICAL REFERENCES

ABCERAM - **Associaçâo Brasileira de Ceràmica** (Electronic document) accessed December 2010. http://www.abceram.org.br

Acchar, W. **Ceramic Materials: Science and Technology**. Natal: EDUFRN, 2000.

Acchar, W.; Ramalho, G.; Fonseca, Y.A.; Hotza, D.; Segadaes, A.M. "Using granite rejects to aid densification and improve mechanical properties of alumina bodies." **Journal of Materials Science**, v. 40, 3905-3909, 2005.

Acchar, W.; Vieira, F.A.; Hotza, D. "Effect of marble and granite sludge in clay materials." **Materials Science and Engineering A**, v. 419, 306309, 2006a.

Acchar, W.; Vieira, F.A.; Segadaes, A.M. "Using ornamental stone cutting rejects as raw materials for red clay ceramic products: Properties and microstructure development." **Materials Science and Engineering A,** v. 435-436, 606-610, 2006b.

Andreola, F.; Barbieri, L.; Corradi, A.; Lancellotti, I.; Manfredini, T. "The possibility to recycle solid residues of the municipal waste incineration into a ceramic tile body." **Journal of Materials Science,** v. 36, 48694873, 2001.

Appendino, P.; Ferraris, M.; Matekovits, I.; Salvo, M. "Production of glassceramics bodies from the bottom ashes of municipal solid waste incinerators." **Journal of the European Ceramic Society,** v. 24, 803810, 2004.

Brazilian Association of Technical Standards: NBR 10004 - **Residuos Sólidos**: Classificaçao, 1987.

Brazilian Association of Technical Standards: NBR 13818 - **Ceramic tiles for cladding**: Specification and test methods, 1997.

Brazilian Association of Technical Standards: NBR 6457 - **Soil Samples**: Preparation for Compaction Tests and Characterization Tests, 1986.

Brazilian Association of Technical Standards: NBR 6459 - **Soil Samples**: Determination of Liquidity Limit, 1984.

Brazilian Association of Technical Standards: NBR 7180 - **Soil Samples**: Determination of the Plasticity Limit, 1984.

Barbieri, L.; Corradi, A.; Lancellotti, T.; Manfredini, T. "Use of municipal incinerator bottom ash as sintering promoter in industrial ceramics." **Waste Management,** v.22, 859-863, 2002.

Bergeron, C.G.; Risbud, S.H. **Introduction to Phase Equilibria in Ceramics.** Westerville, Ohio: American Ceramic Society, 1984.

Boraschi, E.; Cunha J.V.; Vivona, D. Engobes: Characteristics and Applications. Part 1 - A method for evaluating the variation in the degree of waterproofing of engobes with firing temperature. **Cerâmica Industrial,** v.1, n.1, 31-33, 1996.

Callister Jr., W.D. **Materials Science and Engineering: An Introduction**. 5ª edition. Rio de Janeiro: LTC, 1999.

Campos, M.; Velasco, F.; Martinez, M.A.; Torralba, J.M. "Recovered slate waste as raw material for manufacturing sintered structural tiles" **Journal of the European Ceramic Society,** v.24, 811-819, 2004.

Caputo, H.P. **Mecânica dos Solos e suas Aplicações**. v.1, 6ª edição. Rio de Janeiro: LTC, 1998.

Catarino, L.; Sousa, J.; Martins, I.M.; Vieira, M.T.; Oliveira, M.M. "Ceramic products obtained from rock wastes." **Journal of Materials Processing Technology,** v.143-144, 843-845, 2003.

Chiang, Y.M.; Birnie III, D.; Kingery, W.D. **Physical Ceramics: Principles for Ceramic Science and Engineering**. New York: Wiley, 1997.

Colombo, P.; Brusatin, G.; Bernardo, E.; Scarinci, G. "Inertization and reuse of waste materials by vitrification and fabrication of glass-based products." **Current Opinion in Solid State and Materials Science,** v.7, 225-239, 2003.

Conner, J.R. **Environmental Engineering in the Process Plan.** New York: McGraw-Hill, 1992.

Cunha, J.P. "**Development of a new material from the composition of varvito mining waste and lime production.**" Dissertation [Master's Degree in Materials Science and Engineering], Federal University of Paranà, 2007.

Cunha, Y.M. "**Aspectos da paisagem oleira de morro da fumaça (SC)**." Dissertation [Master's in Geography], Federal University of Santa Catarina, 2003.

D'Agostino, L.Z.; Soares, L. "The use of granitic-gneiss rock quarry fines as a substitute for natural sands in mortar production." **Geosciences**, v. 22, 65-73, 2003.

Della, V.P.; Hotza, D.; Kuhn, I. "Characterization of rice husk ash for use as a raw material in the manufacture of silica refractories." **Quimica Nova**, v. 24, 778-782, 2001.

Della, V.P.; Kuhn, I.; Hotza, D. "Rice husk ash as an alternate source for active silica production." **Materials Letters**, v. 57, 818-821, 2002.

Della, V.P.; Kuhn, I.; Hotza, D. "Processing and characterization of active silica obtained from rice husk ash." **Materials Science Forum**, v. 416-418, 531-536, 2003.

Della, V.P.; Kuhn, I.; Hotza, D. "Recycling agro-industrial residues: rice husk ash as an alternative source of silica." **Ceràmica Industrial**, v. 10, 22-25, 2005.

Della, V.P.; Junkes J.A.; Montedo, O.R.K.; Oliveira, A.P.N.; Rambo C.R.; Hotza, D. "Synthesis of Hematite from Steel Scrap to Produce Ceramic Pigments." **American Ceramic Society Bulletin**, v. 86, 9101-9105, 2007.

Dondi, M.; Marsigli, M.; Fabbri, B. Recycling of industrial and urban wastes in brick production - A Review. **Tile & Brick International,** v. 13, 218-225, 1997.

DRM - **Department of Mineral Resources - RJ** (Electronic Document) accessed September 2009. http://www.drm.rj.gov.br

Gregg, S.J.; Sing, K.S.W. "**Adsorption, Surface Area and Porosity.**" Academic Press; London, 1982.

Gonçalves, E.G. **Equilibrium diagrams applied to ceramics**. Course handout offered by the Brazilian Metals Association - ABM, Sao Paulo, 1987.

Handle, F. **Extrusion in Ceramics**. Berlin: Springer, 2007.

Holtz, R.D.; Kovacs, W.D. **An Introduction to Geotechnical Engineering**. New York: Prentice Hall, 1981.

Hoppen, C.; Portella, K.F.; Joukoski, A.; Baron, O.; Franck, R.; Sales, A.; Andreoli, C.V.; Paulon, V.A. "Co-disposition of centrifuged water treatment plant (WTP) sludge in concrete matrix: alternative method for environmental preservation." **Cerâmica**, v.51, 85-95, 2005.

Junior, V.M.T.; Mymrine, V.; Ribeiro, R.A.C.; Ponte, H.A. "Utilization of WTP sludge in the development of new ceramic composites." **17th CBECIMat - Brazilian Congress of Materials Science and Engineering**, Foz do Iguaçu - PR, 2006.

Junkes, J.A.; Della, V.P.; Acchar W.; Oliveira, A.P.N.; Hotza, D. "Obtaining amorphous silica from acid-leached, calcined rice husk." **Industrial Ceramics**, v. 1, 11-15, 2006.

Kingery, W.D.; Bowen, H.K.; Uhlmann, D.R. **Introduction to Ceramics**. 2ª edition. New York: Wiley, 1976.

LaGrega, M.D.; Buckingham, P.L.; Evans, J.C. **Hazardous Waste Management.** New York: McGraw-Hill, 1994.

Levin, E.M., Robbins, C.R., McMurdie, H.F. (editors). **Phase Diagrams for Ceramists.** Columbus, Ohio: American Ceramic Society, 1964, Fig. 407.

Legislaçao - Presidência da Repûblica: Casa Civil - Subchefia para Assuntos Juridicos (Electronic Document) accessed December 2010. http://www.planalto.gov.br/ccivil_03/_ato2007-2010/2010/lei/l12305.htm

Mansur, A.A.P.; Peres, A.E.C.; Palhares, L.; Mansur H.S. "Study of pore size distribution of slate ceramic pieces produced by slip casting of waste powders." **Minerals Engineering,** v.19, 525-527, 2006.

Menezes, R.R.; Neves, G.A.; Ferreira, H.C. "Mapeamento de argilas do Estado da Paraiba." **Ceràmica,** v. 47, 77-81, 2001.

Mitchell, J.K. **Fundamentals of Soil Behavior**. New York: Wiley, 1976.

Modesto, C.; Bristot, V.; Menegali, G.; Brida, M. De; Mazzucco, M.; Mazona, A.; Borba, G.; Virtuoso, J.; Gastaldon, M.; Novaes de Oliveira, A.P. "Obtençao e Caracterizaçao de Materiais Cerâmicos a partir de Residuos Sólidos Industriais". **Ceràmica Industrial**, v.8 n.4, 14-18, 2003.

Monteiro, S.N.; Alexandre, J.; Margem, J.I.; Sànchez, R.; Vieira, C.M.F. "Incorporation of sludge waste from water treatment plant into red ceramic." **Construction and Building Materials**, v.22, 1281-1287, 2007.

Moreira, J.M.S.; Manhaes, J.P.V.T.; Holanda, J.N.F. "Reuse of ornamental rock waste from the northwest of Rio de Janeiro in red ceramics." **Cerâmica,** v.51, 180-186, 2005.

Moreira, J.M.S.; Manhaes, J.P.V.T; Holanda, J.N.F. "Processing of red ceramic using ornamental rock powder waste." **Journal of Materials Processing Technology,** v.196, 88-93, 2008.

Oliveira, E.M.S.; Machado, S.Q.; Holanda, J.N.F. "Caracterizaçao de residuo (sludge) proveniente de estaçao de tratamento de água visando sua utilização em cerâmica vermelha." **Cerâmica**, v.50, 324-330, 2004.

Padilha, A.F. **Engineering Materials: Microstructure and Properties.** Sao Paulo: Hemus, 1997.

Paixao, L.C.C. **"Utilization of water treatment plant sludge in red ceramics."** Dissertation [Master's in Materials Engineering] Federal University of Ouro Preto, 2005.

Pinto, F.A.R. **"Solid industrial waste: characterization and management. The case of the state of Ceará."** Dissertation [Master's in Civil Engineering]. Federal University do Cearâ, 2004.

Pires, J.L. **"The planning of mining activities for the conurbed area of Florianópolis."** Dissertation [Master's in Civil Engineering], Federal University of Santa Catarina, 2000.

Popp, J.H. **General Geology.** 5ª edition. Rio de Janeiro: LTC, 1998.

Quimica **Ambiental** - Environmental **Chemistry Laboratory**, (Electronic Document) accessed March 2010. http://www.usp.br/qambiental/tratamentoAgua.html#tratamento

Raupp-Pereira, F.; Hotza, D.; Segadaes, A.M.; Labrincha, J.A. "Ceramic formulations prepared with industrial wastes and natural by-products." **Ceramics International**, v.32, 173-179, 2006.

Raupp-Pereira, F.; Nunes, A.F.; Segadaes, A.M.; Labrincha, J.A. "Refractory formulations made of different wastes and natural subproducts." **Key Engineering Materials**, v.264-268, 1743-1746, 2004.

Raupp-Pereira, F.; Silva, L.; Segadaes, A.M.; Paiva, H.; Labrincha, J.A. "Effects of potable water filtration sludge on the rheological behavior of one-coat plastering mortars." **Journal of Materials Processing Technology,** v.190, 12-17, 2007.

Reed, J. S. **Principles of Ceramic Processing**. 2ª edition. New York: Wiley, 1995.

Ribeiro, M.J.; Ferreira, A.A.L.; Labrincha, J.A. "Aspectos fundamentais sobre a extrusao de massa de cerâmicas vermelhas." **Cerâmica Industrial**, v.8, 37-42, 2003.

Ring, T.A. **Fundamentals of Ceramic Powder Processing and Synthesis.** New York: Academic Press, 1996.

Rocha, J.C.; John, V.M. **Coletânea Habitare: Utilizaçâo de Residuo na Construçâo Habitacional**, v.4, Porto Alegre: ANTAC, 2003.

SAMAE - Serviço Autònomo Municipal de Agua e Esgoto (Electronic Document) accessed June 2008. http://www.samae.com.br

Santos, P.S. **Ciência e Tecnologia de Argilas**. v.1. Sao Paulo: Edgard Blücher, 1997.

Segadaes, A.M. **Diagrama de Fases - Teoria e Aplicaçâo em Ceràmica.** Sao Paulo: Edgard Blücher, 1987.

Segadaes, A.M. "Use of phase diagrams to guide ceramic production from wastes." **Advances in Applied Ceramics**, v.105, 46-54, 2006.

Segadaes, A.M.; Carvalho, M.A.; Acchar, W. "Using marble and granite rejects to enhance the processing of clay products." **Applied Clay Science**, v.30, 42-52, 2005.

Silva, M.E.M.C.; Peres, A.E.C. "Thermal expansion of slate wastes." **Minerals Engineering** v.19, 518-520, 2006.

Silveira, F.G. **Investigation of the mechanical behavior of a residual gneiss soil from the city of Porto Alegre**. Dissertation [Master's in Civil Engineering]. Federal University of Rio Grande do Sul, Porto Alegre, 2005.

Souza, A.J.; Pinheiro, B.C.A.; Holanda, J.N.F. "Recycling of gneiss rock waste in the manufacture of vitrified floor tiles." **Journal of Environmental Management,** v.91 685-689, 2010.

Souza, L.P. de F.; Mansur, H.S. "Production and characterization of ceramic pieces obtained by slip casting using powder wastes." **Journal of Materials Processing Technology**, v.145, 14-20, 2004.

Souza, J.S. "**Museum and economic development: the case of Varvito Park (Itu, SP_Brazil).**" Museu Paulista - USP. Brazil - Porto Feliz, SP (Electronic Document) accessed June 2008. http://www.naya.org.ar/turismo_cultural/congreso/ponencias/jonas_so ares_de_souza.htm

Teixeira, W.; Toledo, M.C.M.; Fairchild, T.R.; Taioli, F. **Deciphering the Earth**. Sao Paulo: Oficina de Textos, 2000.

Teixeira, S.R.; Souza, S.A.; Souza, N.R. "Feasibility of using waste from water and sewage treatment plants in the ceramics industry." **In: 46th Brazilian Ceramics Congress**, Sao Paulo, 2002a.

Teixeira, S.R.; Souza, S.A.; Souza, N.R.; Job, A.E.; Gomes, H.M.; Heitzmann Neto, J.F. "Characterization of waste from water treatment plants (WTP) and sewage treatment plants (STP) and study of the feasibility of their use by the ceramics industry." **In: XXVIII Congreso Interamericano de Ingeniería Sanitaria y Ambiental**, Cancun, 2002b.

Teixeira, S.R.; Souza, S.A; Souza, N.R.; Aléssio, P.; Santos, G.T.A. "Effect of the addition of water treatment plant (WTP) sludge on the properties of structural ceramic material." **Cerâmica,** v.52, 215220, 2006.

Tomi, G.; Senhorinho, N.C.; Figueiredo Filho, P.M.; Ferrari, K.R.; Chausson, D. "Diagnosis and Actions to Improve the Mining Activities of Raw Materials for the Ceramic Industry." **Cerâmica Industrial**, v.6, 34-36, 2000.

Valente, A.M. and Barbosa, L.M. "The vulnerability of use around mechanized quarries in Feira de Santana, Ba: a model." **X**

Brazilian Symposium on Applied Physical Geography, Rio de Janeiro, 2003.

Van Vlack, L. H. **Properties of Ceramic Materials.** Sao Paulo: Edgard Blücher 1973.

Vieira, J.B. "**Economic and financial evaluation of the implementation of a consortium for the exploitation of clays: a case study in the ceramics production chain in Rondônia.**" Dissertation [Master's in Production Engineering], Federal University of Santa Catarina, 2002.

Vieira, C.M.F.; Henriques, D.N.; Peiter, C.C.; Carvalho, E.A.; Monteiro, S.N. "Utilization of fine gneiss in ceramic mass for roof tiles." **Revista Matéria**, v.11, 211-216, 2006a.

Vieira, C.M.F.; Teixeira, S.S.; Toledo, R.; Souza, S.D.C.; Monteiro, S.N. "Electric porcelain with ornamental rock sawdust residue, part 1: microstructural evolution, physical and mechanical properties." **Revista Matéria**, v.11, 427-434, 2006b.

Villieras, F.; Yvon, J.; Cases, J.M.; Donato, P.De; Lhote, F.; Baeza, R. "Development of microporosity in clinochlore upon heating." **Clays and Clay Minerals**, v.42, 679-688, 1994.

Printed by Books on Demand GmbH, Norderstedt / Germany